Solutions Manual to Accompany

Lectures on the Electrical Properties of Materials (Fourth Edition)

Solutions Manual to Accompany

Lectures on the Electrical Properties of Materials (Fourth Edition)

L. B. Au
and
L. Solymar

Department of Engineering Science,
University of Oxford

Oxford New York Tokyo
OXFORD UNIVERSITY PRESS
1988

Oxford University Press, Walton Street, Oxford OX2 6DP
Oxford New York Toronto
Delhi Bombay Calcutta Madras Karachi
Petaling Jaya Singapore Hong Kong Tokyo
Nairobi Dar es Salaam Cape Town
Melbourne Auckland
and associated companies in
Berlin Ibadan

Oxford is a trade mark of Oxford University Press

Published in the United States
by Oxford University Press, New York

British Library Cataloguing in Publication Data
Au, L. B.
Solutions manual to accompany lectures
on the electrical properties of materials
1. Materials—Electric properties
I. Title. II. Solymar, L
620.1'1297 QC522
ISBN 0–19–856221–7

Printed in Great Britain
by Biddles Ltd
Guildford & King's Lynn

Contents

1. The electron as a particle

1.1 Electric field $\mathscr{E} = \dfrac{\text{Potential difference}}{\text{length}} = \dfrac{10\times10^{-3}}{10\times10^{-3}}\ \text{Vm}^{-1} = 1\ \text{Vm}^{-1}$

Current density $J = \dfrac{\text{current}}{\text{area}} = \dfrac{6.4\times10^{-3}}{(10\times10^{-3})^2}\ \text{Am}^{-2} = 64\ \text{Am}^{-2}$

Since $J = \sigma\mathscr{E}$ [eqn 1.24] where σ is the electrical conductivity,

$$\sigma = J/\mathscr{E} = 64/1 = 64\ \Omega^{-1}\text{m}^{-1}$$

From [eqn 1.10] : $\sigma = N_e e\mu_e$ where $\mu_e = e\tau/m$

$\therefore$ density of carriers, $N_e = \dfrac{\sigma}{e\mu_e} = \dfrac{64}{1.6\times10^{-19}\times0.39} = \underline{1.03\times10^{21}\ \text{m}^{-3}}$

Replace m by m^*, the effective mass, in the expression of μ_e, we have

then the collision time of carriers, $\tau = \dfrac{m^*\mu_e}{e} = \dfrac{0.12\times9.11\times10^{-31}\times0.39}{1.6\times10^{-19}}$

$$= \underline{2.66\times10^{-13}\ \text{s}}$$

1.2 (i) The angular frequency of the propagating wave is given by

$$\omega = 2\pi f = 2\pi c/\lambda = \frac{2\pi\times3\times10^8}{0.5\times10^{-3}} = 3.77\times10^{12}\ \text{rad s}^{-1}$$

There is a resonant absorption when the frequency of the wave is equal to the cyclotron frequency,

i.e. $\omega = \omega_c = eB/m^*$ [eqn 1.68]

Hence $m^* = eB/\omega = \dfrac{1.6\times10^{-19}\times0.323}{3.77\times10^{12}} = 1.37\times10^{-32}$ kg

$= \underline{0.015\ m_o}$ where m_o is the free electron mass

(ii) Collision time of InSb = 15 x collision time of Ge = 4.0×10^{-12} s

$$\therefore \text{ mobility, } \mu_e = \frac{e\tau}{m^*} = \frac{1.6\times10^{-19}\times4.0\times10^{-12}}{1.37\times10^{-32}} = \underline{46.7\ m^2V^{-1}s^{-1}}$$

(iii) The sharpness of resonance is determined by $\omega_c\tau$.

Now $\omega_c\tau = 3.77\times10^{12}\times4.0\times10^{-12} \simeq 15 \gg 1$; therefore, the resonance may be regarded as sharp.

1.3 Conductivity $\sigma = \sigma_e + \sigma_h = N_e e\mu_e + N_h e\mu_h$

But $\mu_e = 10\ \mu_h$, therefore $\sigma = e\mu_h(10N_e + N_h)$

Put in the numerical values,

$$0.455 = 1.6\times10^{-19}\mu_h(10\times10^{19} + 10^{20})$$

yielding $\mu_h = \underline{0.014\ m^2V^{-1}s^{-1}}$ and $\mu_e = \underline{0.14\ m^2V^{-1}s^{-1}}$

1.4 (i) Since electric conduction in sodium is caused by electrons only, the Hall coefficient, $R = \dfrac{1}{N_e e}$ [eqn 1.20].

$$\therefore N_e = 1/Re = 1/(2.5\times10^{-10}\times1.6\times10^{-19}) = \underline{2.5\times10^{28}\ m^{-3}}$$

(ii) Since $\sigma = N_e e\mu_e = \mu_e/R$,

$$\mu_e = \sigma R = 2.5\times10^{-10}/4.7\times10^{-8} = \underline{5.32\times10^{-3}\ m^2V^{-1}s^{-1}}$$

(iii) The material is critically transparent when the frequency of the wave is equal to the plasma frequency, i.e. $\omega^2 = \omega_p^{\ 2} = \dfrac{N_e e^2}{m^* \varepsilon_o}$ (from [eqn 1.53])

whence $m^* = \dfrac{N_e e^2}{\omega^2 \varepsilon_o} = N_e e^2 / \left[\dfrac{2\pi c}{\lambda}\right]^2 \varepsilon_o = \dfrac{2.5\text{x}10^{28}\text{x}(1.6\text{x}10^{-19})^2}{\left(\dfrac{2\pi\text{x}3\text{x}10^8}{210\text{x}10^{-9}}\right)^2 \text{x}8.85\text{x}10^{-12}}$

$= 8.976\text{x}10^{-31}$ kg $= \underline{0.99m_o}$

(iv) Since $\mu_e = e\tau/m^*$, $\tau = m^*\mu_e/e = \dfrac{8.976\text{x}10^{-31}\text{x}5.32\text{x}10^{-3}}{1.6\text{x}10^{-19}} = \underline{2.98\text{x}10^{-14}\text{ s}}$

(v) $6.02\text{x}10^{26}$ Na atoms have a mass of 23 kg. Now density = 971 kgm^{-3}, therefore the number of atoms per m^3 is

$$N_a = \frac{6.02\text{x}10^{26}}{23} \text{ x } 971 = 2.54\text{x}10^{28}\ \text{m}^{-3}$$

Hence the number of electrons per atom available for conduction is

$$\frac{N_e}{N_a} = \frac{2.5\text{x}10^{28}}{2.54\text{x}10^{28}} = \underline{0.98}$$

1.5 The wave number k is given by [eqn 1.51] :

$$k = \omega\sqrt{\mu\varepsilon}\left(1 + \frac{i\sigma}{\omega\varepsilon(1 - i\omega\tau)}\right)^{1/2} \qquad (1)$$

The real and imaginary parts of k for $\sigma = 1/4.8\text{x}10^{-8}$ Ωm, $\tau = 2.98\text{x}10^{-14}$ s (values obtained from Ex 1.4) are plotted in Fig 1.1.

The penetration depth δ is given by

Im{k} δ = 1 where k is defined in eqn (1)

For frequency = 10^6, 10^{15}, $2\text{x}10^{15}$ Hz, the corresponding penetration depths are $\underline{1.09\text{x}10^{-4},\ 4.67\text{x}10^{-8},\ 2.45\text{x}10^{-5}\text{ m}}$ respectively.

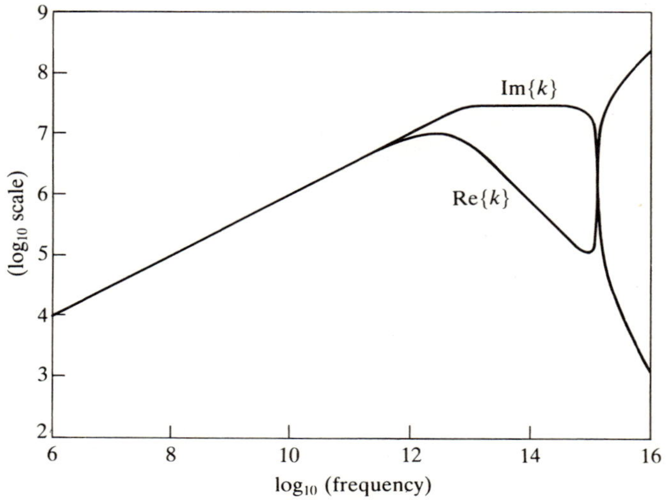

Fig 1.1 Plot of the real and imaginary parts of k as a function of frequency.

1.6 (i) There may be a misalignment of contacts.

The potential difference across the 5mm length = 310 mV

Now there is a 3.2 mV measured, therefore the misaligned distance

$$\approx \frac{3.2}{310} \times 5 \approx \underline{0.05 \text{ mm}}$$

(ii) Corrected Hall voltage = 8.0 - 3.2 = 4.8 mV

Using [eqn 1.20] :

$$N = \frac{JB}{e\mathscr{E}_H} = \frac{(5\times10^{-3}/2\times10^{-6})\times0.16}{1.6\times10^{-19}\times(4.8/2)} = \underline{1.04\times10^{21}\ \text{m}^{-3}}$$

(iii) Since $J = \sigma\mathscr{E}$,

$$5\times10^{-3}/2\times10^{-6} = \sigma\ (310/5) \quad \text{(when } B = 0\text{), giving} \quad \sigma = \underline{40.3\ \Omega^{-1}\text{m}^{-1}}$$

(iv) From results of parts (ii) and (iii), and since $\sigma = Ne\mu$,

$$\mu = 40.3/(1.04\times10^{21}\times1.6\times10^{-19}) = \underline{0.24\ \text{m}^2\text{V}^{-1}\text{s}^{-1}}$$

1.7 No, because the recombinations of electrons and holes will prevent an infinite accumulation of both charges. A dynamic equilibrium will be reached when the rate of recombination is equal to the rate of supply of the charges.

1.8 In steady state (i.e. $\frac{\partial}{\partial t} = 0$), the equations of motion for electrons and holes are

$$\begin{cases} \dfrac{m_e \mathbf{v_e}}{\tau_e} = -e(\boldsymbol{\mathscr{E}} + \mathbf{v_e} \times \mathbf{B}) \\[2ex] \dfrac{m_h \mathbf{v_h}}{\tau_h} = e(\boldsymbol{\mathscr{E}} + \mathbf{v_h} \times \mathbf{B}) \end{cases}$$

where $\mathbf{v_e}$, $\mathbf{v_h}$, $\boldsymbol{\mathscr{E}}$, $\mathbf{B}$ are the velocities of electrons, holes, the electric field (including both the applied field and Hall field), and the magnetic field respectively.

Since B-field is in the y-direction (refer to the coordinate system of [Fig 1.3]) and the $\mathscr{E}$-field has components in the x and z directions only, the charge carriers will be moving in the x-z plane. Thus

$$\begin{cases} v_{ex}\mathbf{i} + v_{ez}\mathbf{k} = -\dfrac{e\tau_e}{m_e}\left[\mathscr{E}_H\mathbf{i} - \mathscr{E}_a\mathbf{k} - v_{ez}B_y\mathbf{i} + v_{ex}B_y\mathbf{k}\right] \\[2ex] v_{hx}\mathbf{i} + v_{hz}\mathbf{k} = \dfrac{e\tau_h}{m_h}\left[\mathscr{E}_H\mathbf{i} - \mathscr{E}_a\mathbf{k} - v_{hz}B_y\mathbf{i} + v_{hx}B_y\mathbf{k}\right] \end{cases}$$

Since $v_{ex}B_y$ and $v_{hx}B_y$ are small compared with $\mathscr{E}_a$, the equations reduce to

$$\begin{cases} v_{ex} = -\mu_e(\mathscr{E}_H - v_{ez}B_y) \\ v_{ez} = \mu_e\mathscr{E}_a \\ v_{hx} = \mu_h(\mathscr{E}_H - v_{hz}B_y) \\ v_{hz} = -\mu_h\mathscr{E}_a \end{cases} \tag{1}$$

The longitudinal current $J_z = -N_e e v_{ez} + N_h e v_{hz}$

$$= -(N_e\mu_e + N_h\mu_h)e\mathscr{E}_a \qquad (2)$$

There is no transverse circuit, so the transverse current must be zero,

i.e. $J_x = 0$

or $-N_e e v_{ex} + N_h e v_{hx} = 0$

With eqn (1), it becomes

$$N_e e\mu_e(\mathscr{E}_H - \mu_e\mathscr{E}_a B_y) + N_h e\mu_h(\mathscr{E}_H + \mu_h\mathscr{E}_a B_y) = 0$$

$$\mathscr{E}_H = \frac{N_e\mu_e^2 - N_h\mu_h^2}{N_e\mu_e + N_h\mu_h}\,\mathscr{E}_a B_y$$

$$= -\frac{N_e\mu_e^2 - N_h\mu_h^2}{(N_e\mu_e + N_h\mu_h)^2}\cdot\frac{1}{e}\cdot J_z B_y \qquad \text{(from eqn (2))}$$

Thus, by the definition of the Hall coefficient [eqn 1.20],

$$R = -\frac{N_e\mu_e^2 - N_h\mu_h^2}{e\,(N_e\mu_e + N_h\mu_h)^2}$$

We can see that if $\mu_e >> \mu_h$, the Hall coefficient may be negative though

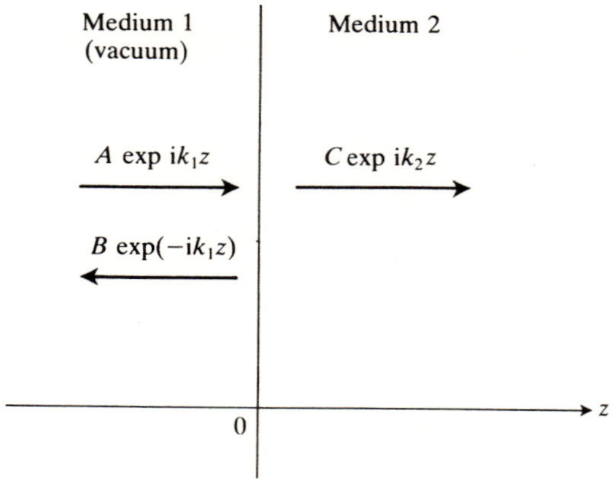

Fig 1.2 Incident, transmitted and reflected waves.

$N_e < N_h$. Therefore a negative Hall coefficient does not necessarily mean that electrons are the dominant charge carriers.

1.9 Choose a coordinate system as in Fig 1.2. Suppose that the electric fields have components in the x direction only. Let the waves be described by

$$\begin{cases} \mathscr{E}_{1x} = A \exp ik_1 z + B \exp(-ik_1 z) \\ \mathscr{E}_{2x} = C \exp ik_2 z \end{cases}$$

where the subscripts 1, 2 represent medium 1 and medium 2. Constants A, B, and C describe the amplitudes of the incident, reflected, and transmitted waves respectively. The wave numbers are given by

$$\begin{cases} k_1 = \omega\sqrt{\mu_o \varepsilon_o} \quad \text{(medium 1 is vacuum)} \\ k_2 = \omega\sqrt{\mu_o \varepsilon_o} \left(1 - \omega_p^2/\omega^2\right)^{1/2} \qquad \text{[eqn 1.54]} \end{cases}$$

From Maxwell's equation : curl $\boldsymbol{\mathscr{E}} = -\dot{\mathbf{B}}$

Note that $\frac{\partial}{\partial t} = -i\omega$. Therefore

$$\begin{cases} H_{1y} = \dfrac{k_1}{\omega\mu_o} \left[A \exp ik_1 z - B \exp(-ik_1 z) \right] \\ H_{2y} = \dfrac{k_2}{\omega\mu_o} C \exp ik_2 z \end{cases}$$

At $z = 0$ (i.e. at the boundary), the electric fields and magnetic fields must be the same in both mediums. Therefore

$$\begin{cases} A + B = C \\ \dfrac{k_1}{\omega\mu_o} (A - B) = \dfrac{k_2}{\omega\mu_o} C \end{cases}$$

giving $\frac{B}{A} = \frac{k_1 - k_2}{k_1 + k_2}$ and $\frac{C}{A} = \frac{2k_1}{k_1 + k_2}$

Now power of a wave $= \frac{1}{2} \left| \mathrm{Re}\{ \mathcal{E} \times \mathbf{H}^* \} \right|$. Therefore

$$\left\{\begin{aligned} &\text{incident power} = \frac{k_1}{2\omega\mu_0} |A|^2 \\ &\text{reflected power} = \frac{k_1}{2\omega\mu_0} |B|^2 \\ &\text{transmitted power} = \frac{|C|^2}{2\omega\mu_0} \mathrm{Re}\{k_2^*\} = \frac{|C|^2}{2\omega\mu_0} \mathrm{Re}\{k_2\} \end{aligned}\right.$$

Hence amount of reflection $= \frac{\text{reflected power}}{\text{incident power}} = \left|\frac{B}{A}\right|^2 = \left|\frac{k_1 - k_2}{k_1 + k_2}\right|^2$

$$= \left| \frac{1 - (1 - \omega_p^2/\omega^2)^{1/2}}{1 + (1 - \omega_p^2/\omega^2)^{1/2}} \right|^2$$

When $\omega_p < \omega$, k_2 is real and the amount of transmission

$$= \frac{\text{transmitted power}}{\text{incident power}} = \frac{k_2}{k_1} \left|\frac{C}{A}\right|^2 = \frac{4k_1k_2}{(k_1 + k_2)^2} = \frac{4(1 - \omega_p^2/\omega^2)^{1/2}}{\left[1 + (1 - \omega_p^2/\omega^2)^{1/2} \right]^2}$$

However, when $\omega_p > \omega$, k_2 is imaginary, its real part is zero, and the transmitted power is zero. It is interesting to know that in this case $\left|\frac{B}{A}\right|^2 = 1$, i.e. all the power is reflected.

1.10 Use the same notations as in Ex 1.9. The boundary between the second medium and the third medium is at $z = d$.

In medium 1 (vacuum),

$$\begin{cases} \mathscr{E}_{1x} = A \exp ik_1 z + B \exp(-ik_1 z) \\ H_{1y} = \dfrac{k_1}{\omega\mu_o}\left[A \exp ik_1 z - B \exp(-ik_1 z) \right] \end{cases}$$

where $k_1 = \omega\sqrt{\mu_o \varepsilon_o}$

In medium 2,

$$\begin{cases} \mathscr{E}_{2x} = C \exp ik_2 z + D \exp(-ik_2 z) \\ H_{2y} = \dfrac{k_2}{\omega\mu_o}\left[C \exp ik_2 z - D \exp(-ik_2 z) \right] \end{cases}$$

where $k_2 = \omega\sqrt{\mu_o \varepsilon_o}\left(1 - \omega_p^2/\omega^2\right)^{1/2} = \left(\omega^2 - \omega_p^2\right)^{1/2}/c$, c is the speed of light in vacuum.

In medium 3 (vacuum; note that $k_3 = k_1$),

$$\begin{cases} \mathscr{E}_{3x} = F \exp ik_1(z - d) \\ H_{3y} = \dfrac{k_1}{\omega\mu_o} F \exp ik_1(z - d) \end{cases}$$

Matching the electric and magnetic fields at $z = 0$ gives

$$\begin{cases} A + B = C + D \\ k_1(A - B) = k_2(C - D) \end{cases} \tag{1}$$

Similarly, matching at $z = d$ gives

$$\begin{cases} C \exp ik_2 d + D \exp(-ik_2 d) = F \\ k_2\left[C \exp ik_2 d - D \exp(-ik_2 d) \right] = k_1 F \end{cases} \tag{2}$$

Eliminate B from eqn (1), we get

$$2A = (1 + k_2/k_1)C + (1 - k_2/k_1)D \tag{3}$$

From eqn (2),

$$\begin{cases} 2C \exp ik_2 d = (1 + k_1/k_2)F \\ 2D \exp(-ik_2 d) = (1 - k_1/k_2)F \end{cases}$$

Substitute into eqn (3),

$$\frac{4A}{F} = (1+k_2/k_1)(1+k_1/k_2)\exp(-ik_2d) + (1-k_2/k_1)(1-k_1/k_2)\exp ik_2d$$

$$= 4\cosh(k_{2i}d) + 2i\left(\frac{k_{2i}}{k_1} - \frac{k_1}{k_{2i}}\right)\sinh(k_{2i}d) \quad \text{where } k_{2i} = ik_2$$

Thus the ratio of the transmitted to incident power

$$= \left|\frac{F}{A}\right|^2 = \left\{\cosh^2(k_{2i}d) + \frac{1}{4}\left(\frac{k_{2i}}{k_1} - \frac{k_1}{k_{2i}}\right)^2\sinh^2(k_{2i}d)\right\}^{-1}$$ (because

in our case, $\omega_p > \omega$ which means that k_{2i} is real)

$$= \left\{\cosh^2\left[\left(\omega_p^2 - \omega^2\right)^{1/2}\frac{d}{c}\right] + \frac{1}{4}\left[\left(\frac{\omega_p^2}{\omega^2} - 1\right)^{1/2} - \left(\frac{\omega_p^2}{\omega^2} - 1\right)^{-1/2}\right]^2 \sinh^2\left[\left(\omega_p^2 - \omega^2\right)^{1/2}\frac{d}{c}\right]\right\}^{-1}$$

As $\omega\tau \gg 1$, no power is being absorbed.

incident power = transmitted power + reflected power

Thus $\dfrac{\text{reflected power}}{\text{incident power}} = 1 - \dfrac{\text{transmitted power}}{\text{incident power}}$

For $\omega = 6.28\times10^{15}$ rad s^{-1}, $\omega_p = 9\times10^{15}$ rad s^{-1}, $d = 0.25$, 2.5 μm,

$$\left(\omega_p^2 - \omega^2\right)^{1/2}\frac{d}{c} \gg 1 \quad \text{and} \quad \left[\left(\frac{\omega_p^2}{\omega^2} - 1\right)^{1/2} - \left(\frac{\omega_p^2}{\omega^2} - 1\right)^{-1/2}\right]^2 \ll 1$$

Therefore amount of transmission $= \left|\dfrac{F}{A}\right|^2 \simeq 4\exp\left[-\dfrac{2d}{c}\left(\omega_p^2 - \omega^2\right)^{1/2}\right]$

$$= \begin{cases} 8.62\times10^{-5} & \text{for } d = 0.25\ \mu\text{m} \\ 8.67\times10^{-47} & \text{for } d = 2.5\ \mu\text{m} \end{cases}$$

1.11 $\mathbf{J}_{total} = \mathbf{J} + j\omega\varepsilon \boldsymbol{\mathscr{E}}$ (1)

For electrons,

$$\mathbf{J}_e = - N_e e \mathbf{v}_e \text{ , and} \tag{2}$$

$$m \frac{\partial \mathbf{v}_e}{\partial t} = - e \, (\boldsymbol{\mathscr{E}} + \mathbf{v}_e x \mathbf{B}) \tag{3}$$

Replace $\frac{\partial}{\partial t}$ by $-i\omega$ in eqn (3), we have

$$i\omega m_e \mathbf{v}_e / e = \boldsymbol{\mathscr{E}} + \mathbf{v}_e x \mathbf{B}$$

Equate each components

$$\begin{cases} i\omega m_e v_{ex}/e = \mathscr{E}_x + v_{ey} B_o \\ i\omega m_e v_{ey}/e = \mathscr{E}_y - v_{ex} B_o \\ i\omega m_e v_{ez}/e = \mathscr{E}_z \end{cases}$$

Solving for the velocities

$$\begin{cases} v_{ex} = (i\omega m_e \mathscr{E}_x/e + B_o \mathscr{E}_y)/S_e \\ v_{ey} = (i\omega m_e \mathscr{E}_y/e - B_o \mathscr{E}_x)/S_e \\ v_{ez} = e\mathscr{E}_z/i\omega m_e \end{cases} \tag{4}$$

where $S_e = B_o^2 - (m_e \omega/e)^2$

Similarly, for the positively charged ions,

$$\mathbf{J}_i = N_i e \mathbf{v}_i \text{ , and} \tag{5}$$

$$\begin{cases} v_{ix} = (-i\omega m_i \mathscr{E}_x/e + B_o \mathscr{E}_y)/S_i \\ v_{iy} = (-i\omega m_i \mathscr{E}_y/e - B_o \mathscr{E}_x)/S_i \\ v_{iz} = -e\mathscr{E}_z/\omega m_i \end{cases} \tag{6}$$

where $S_i = B_o^2 - (m_i \omega/e)^2$

Eliminate **J** from eqns (1), (2), (5) and rearrange (note that $\mathbf{J} = \mathbf{J}_e + \mathbf{J}_h$)

$$\mathbf{J}_{total} = i\omega \begin{pmatrix} -\dfrac{N_e e v_{ex}}{i\omega} + \dfrac{N_i e v_{ix}}{i\omega} + \varepsilon\mathscr{E}_x \\[2ex] -\dfrac{N_e e v_{ey}}{i\omega} + \dfrac{N_i e v_{iy}}{i\omega} + \varepsilon\mathscr{E}_y \\[2ex] -\dfrac{N_e e v_{ez}}{i\omega} + \dfrac{N_i e v_{iz}}{i\omega} + \varepsilon\mathscr{E}_z \end{pmatrix}$$

With eqns (4) and (6), we can write $\mathbf{J}_{total} = i\omega\bar{\bar{\varepsilon}}_{eqv}\boldsymbol{\mathscr{E}}$ where

$$\bar{\bar{\varepsilon}}_{eqv} = \begin{bmatrix} -\dfrac{N_e m_e}{S_e} - \dfrac{N_i m_i}{S_i} + \varepsilon & -\dfrac{iB_o e}{\omega}\left(-\dfrac{N_e}{S_e} + \dfrac{N_i}{S_i}\right) & 0 \\[2ex] \dfrac{iB_o e}{\omega}\left(-\dfrac{N_e}{S_e} + \dfrac{N_i}{S_i}\right) & -\dfrac{N_e m_e}{S_e} - \dfrac{N_i m_i}{S_i} + \varepsilon & 0 \\[2ex] 0 & 0 & \dfrac{N_e e^2}{m_e \omega^2} + \dfrac{N_i e^2}{m_i \omega^2} + \varepsilon \end{bmatrix}$$

2. The electron as a wave

2.1 The average thermal velocity is given by : $\frac{3}{2}kT = \frac{1}{2}mv^2$

de Broglie relation [eqn 2.6] : $\lambda = h/mv$

Eliminate v to get $\lambda = h/\sqrt{3kTm}$. Hence,

(i) $\lambda = 6.62\text{x}10^{-34}/\left(3\text{x}1.38\text{x}10^{-23}\text{x}300\text{x}9.11\text{x}10^{-31}a\right)^{1/2}$

$= \underline{6.22\ \text{x}10^{-9}/\sqrt{a}\ \text{m}}$

(ii) Mass of a proton is $1.67\text{x}10^{-27}$ kg

$\therefore$ mass of a helium atom $= 4\text{x}1.67\text{x}10^{-27} = 6.68\text{x}10^{-27}$ kg (note that mass of electrons is negligible).

So $\lambda = 6.62\text{x}10^{-34}/\left(3\text{x}1.38\text{x}10^{-23}\text{x}300\text{x}6.68\text{x}10^{-27}\right)^{1/2} = \underline{7.27\text{x}10^{-11}\ \text{m}}$

(iii) Mass of an α-particle (contains four protons) $= 6.68\text{x}10^{-27}$ kg

$\lambda = h/mv = h/\sqrt{2Em}$ because kinetic energy $E = \frac{1}{2}mv^2$

$= 6.62\text{x}10^{-34}/\left(2\text{x}10\text{x}10^{6}\text{x}1.6\text{x}10^{-19}\text{x}6.68\text{x}10^{-27}\right)^{1/2} = \underline{4.53\text{x}10^{-15}\ \text{m}}$

2.2 The spectral composition of the wave packet is given by its Fourier transform,

i.e. $F(\omega) = \int_{-\infty}^{\infty} f(u)\, e^{i\omega t}\, dt$

At $x = 0$, $f(u) = \begin{cases} e^{i\Omega t} & \text{for } |t| < u_o/\Omega \\ 0 & \text{otherwise} \end{cases}$

$$F(\omega) = \int_{-u_o/\Omega}^{u_o/\Omega} e^{i\Omega t}\, e^{i\omega t}\, dt = \frac{2\sin\left(\frac{\Omega+\omega}{\Omega}u_o\right)}{\Omega+\omega}$$

The maximum value of it is $2u_o/\Omega$ when $\omega = -\Omega$. It drops to 63% of this value when $\frac{\Omega+\omega}{\Omega}u_o = \frac{\pi}{2}$.

$\therefore$ the spectral width $= \pi\Omega/u_o$

Momentum $p = h/\lambda$ (deduce from de Broglie's relationship) $= h\omega/2\pi v$

$$\therefore\ \Delta p\, \Delta \ell = \frac{h}{2\pi v}\Delta\omega\, \Delta\ell = \frac{h}{2\pi v}\frac{\pi\Omega}{u_o}\frac{2u_o v}{\Omega} = h \qquad \text{(Q.E.D.)}$$

2.3 (i) Kinetic energy gained by electron

= charge x potential difference = 50 keV

Thus the de Broglie wavelength is

$$\lambda = h/\sqrt{2Em} = 6.62\times10^{-34}/\left(2\times50\times10^{3}\times1.6\times10^{-19}\times9.11\times10^{-31}\right)^{1/2}$$

$$= 5.48\times10^{-12}\ \text{m}$$

Resolution is of the order of the wavelength, i.e. about $\underline{5\times10^{-12}\ \text{m}}$.

(ii) For the same resolution, the wavelength must be the same. Therefore Em = constant. Now mass ratio of neutron to electron is about 1835.

$$\therefore\ E_{neutron} = \frac{50}{1835}\ \text{keV} = \underline{27.2\ \text{eV}}$$

(iii) Lens aberrations, voltage stability etc.

2.4 Since the electrons have low incident energies, scattering takes place predominantly from the surface planes. The grating equation is, therefore, $d \sin \theta = n\lambda$ for maximum intensities (see Fig 2.1).

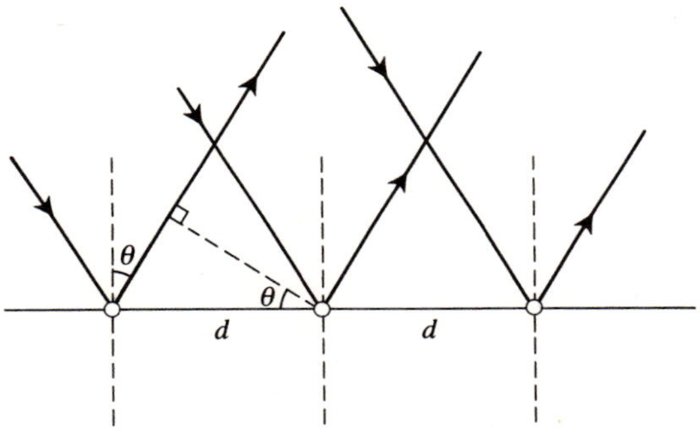

Fig 2.1 Scattering of the incident wave by the surface plane of the crystal.

For E = 70 eV,

$$\lambda = h/\sqrt{2Em} = 6.62\text{x}10^{-34}/\left(2\text{x}70\text{x}1.6\text{x}10^{-19}\text{x}9.11\text{x}10^{-31} \right)^{1/2} = 1.47\text{x}10^{-10}\ \text{m}$$

Hence the intensities are maximum at $\theta = \sin^{-1}(n\lambda/d) = \sin^{-1}\left(\frac{1.47n}{3.52}\right)$

i.e. at $\theta = 0$, $\sin^{-1}0.42$ and $\sin^{-1}0.83$ (for n = 0, 1, 2)

If the crystal is perfect, the maxima are very sharp, and the scattering pattern would look like Fig 2.2.

2.5 There will be maxima whenever a large enough number of atoms radiate in phase. Thus we shall consider only the angles in the two principal planes ($\phi = 0$, $\phi = \pi/2$; refer to Fig. 2.3) and in the diagonal

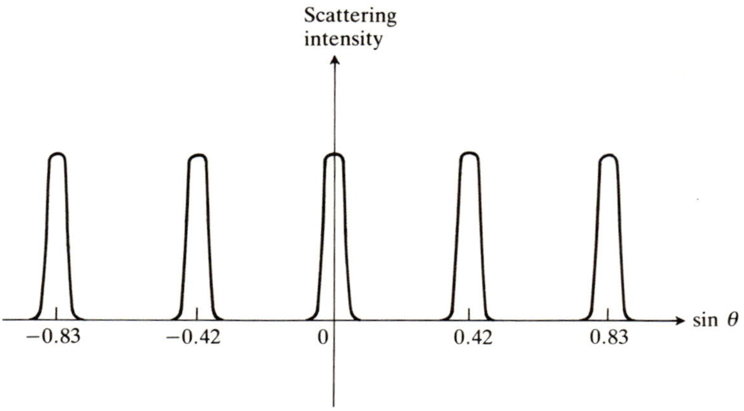

Fig 2.2 Scattering intensity as a function of angles.

plane ($\phi = \pi/4$), in which the number of atoms are large.

For $\phi = 0$ and $\phi = \pi/2$, the interatomic distance is d; therefore the grating equation is $\sin\theta = n\lambda/d$.

For $\phi = \pi/4$, the interatomic distance is $\sqrt{2}\,d$, so the grating equation is $\sin\theta = n\lambda/\sqrt{2}\,d$.

At 10 keV, $\lambda = h/\sqrt{2Em} = 1.23\times10^{-11}$ m

Since $\lambda \ll d$, the first maxima is at a small angle. Hence the spots on

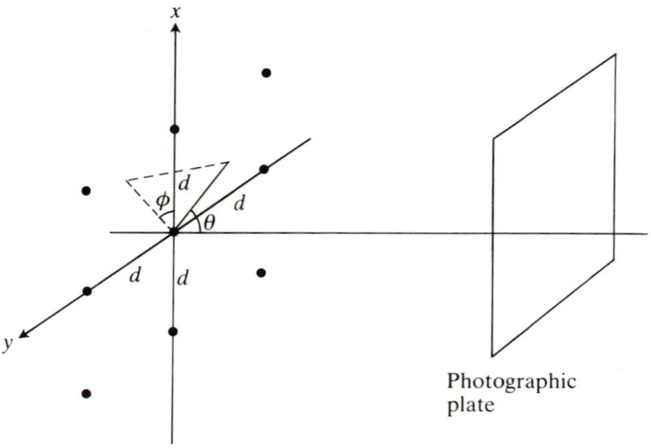

Fig 2.3 Scattering of the wave by a thin foil of crystal.

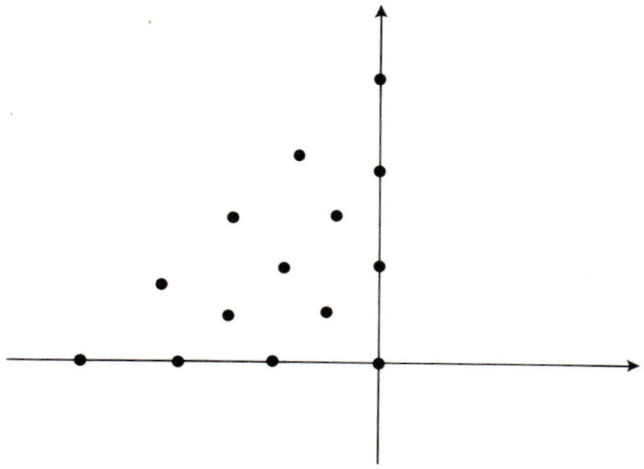

Fig 2.4 Diffraction pattern on the photographic plate.

the photographic plate will be at

$$x_n = y_n \simeq 0.1\ \theta_n \simeq 0.1 \times \frac{1.23\times10^{-11}}{3.52\times10^{-10}}\ n = 3.48\times10^{-3} n$$

In the diagonal planes, the distance between the spots is $\sqrt{2}$ times smaller. Thus the photographic plate will look like Fig. 2.4.

When the material is polycrystalline, one may expect to find the principal planes at any azimuthal direction. Hence the diffraction :n pattern will consist of a set of rings.

2.6 From [eqn 1.77] : $\hbar\omega_p = \hbar\left(\frac{N_e e^2}{m\varepsilon_o}\right)^{1/2} = e\hbar\left(\frac{N_e}{m\varepsilon_o}\right)^{1/2}$

No. of atoms in 1 m^3 of the material $= N_a = \frac{6.02\times10^{26}}{27} \times 2700$

Now each atom contributes three free electrons,

$\therefore N_e = 3N_a = 1.806\times10^{29}\ m^{-3}$

So $\omega_p = \frac{6.62\times10^{-34}}{2\pi} \times \left(\frac{1.806\times10^{29}}{9.11\times10^{-31}\times8.85\times10^{-12}}\right)^{1/2}$ eV $=$ $\underline{15.8\ \text{eV}}$

which is similar to the characteristic energy loss found in the experiment.

3. The electron

3.1 The uncertainty in position is $\Delta x = 2a$.

$\therefore$ according to [eqn 3.44], the uncertainty in momentum $\Delta p = h/2a$.

So p varies between 0 and h/2a. Its mean value is h/4a and the corresponding energy $E = \frac{h^2}{8m(2a)^2}$ which is in agreement with [eqn 3.42] for n = 1.

3.2 For anti-symmetric wavefunction, i.e. $\psi = A \sin kz$ [eqn 3.31] where $k = \frac{n\pi}{2a}$ (from [eqn 3.41] where we take L = 2a),

$$\int_{-a}^{a} |\psi|^2 \, dz = \int_{-a}^{a} |A|^2 \sin^2 kz \, dz = |A|^2 a$$

$$\text{So} \quad \langle z \rangle = \int_{-a}^{a} \psi^* z \psi \, dz \Big/ \int_{-a}^{a} |\psi|^2 \, dz = \frac{1}{a} \int_{-a}^{a} z \sin^2 kz \, dz$$

$$= 0 \quad \text{(because the integrand is an odd function)}$$

$$\langle (z - \langle z \rangle)^2 \rangle = \langle z^2 \rangle = \frac{1}{a} \int_{-a}^{a} z^2 \sin^2 kz \, dz = \frac{1}{a} \int_{-a}^{a} z^2 \, \frac{1 - \cos 2kz}{2} \, dz$$

$$= \frac{a^2}{3} - \frac{1}{k^3 a} \int_{0}^{ka} t^2 \cos 2t \, dt \quad \text{(put } t = kz\text{)}$$

$$= \frac{a^2}{3} - \frac{8a^2}{n^3 \pi^3} \left\{ \frac{t^2 \sin 2t}{2} \Bigg|_{0}^{ka} - \int_{0}^{ka} t \sin 2t \, dt \right\}$$

$$= \frac{a^2}{3} - \frac{8a^2}{n^3\pi^3}\left[\frac{t\cos 2t}{2} - \frac{\sin 2t}{4}\right]_0^{n\pi/2}$$

$$= \frac{a^2}{3}\left(1 - (-1)^n \frac{6}{n^2\pi^2}\right)$$

$$\langle p \rangle = \int_{-a}^{a} \psi^* \left(-i\hbar\frac{\partial}{\partial z}\right) \psi \, dz \Big/ \int_{-a}^{a} |\psi|^2 \, dz$$

$$= -\frac{i\hbar}{a}\int_{-a}^{a} \sin\left(\frac{n\pi}{2a} z\right) \cdot \frac{n\pi}{2a}\cos\left(\frac{n\pi}{2a} z\right) dz$$

$$= -\frac{i\hbar n\pi}{4a^2}\int_{-a}^{a} \sin\left(\frac{n\pi}{a} z\right) dz$$

$= 0$ (because the integrand is odd)

$$\langle (p - \langle p \rangle)^2 \rangle = \langle p^2 \rangle$$

$$= -\frac{\hbar^2}{a}\int_{-a}^{a} \sin\left(\frac{n\pi}{2a} z\right) \cdot \frac{\partial^2}{\partial z^2}\left(\sin\left(\frac{n\pi}{2a} z\right)\right) dz$$

$$= \frac{\hbar^2 n^2 \pi^2}{4a^3}\int_{-a}^{a} \sin^2\left(\frac{n\pi}{2a} z\right) dz$$

$$= \hbar^2 \left(\frac{n\pi}{2a}\right)^2 \quad (= \hbar^2 k_n^2 = 2mE_n)$$

For symmetric wavefunctions, the calculations are similar and the results are the same except $\langle (z - \langle z \rangle)^2 \rangle = \frac{a^2}{3}\left(1 + (-1)^n \frac{6}{n^2\pi^2}\right)$

3.3 (i) Classically, there is an equal probability for the particle to be anywhere between $\pm$ a. Hence $\langle z \rangle = 0$. Similarly $\langle p \rangle = 0$.

$$\therefore \langle (z - \langle z\rangle)^2 \rangle = \langle z^2 \rangle = \frac{1}{2a}\int_{-a}^{a} z^2 \, dz = \frac{a^2}{3}$$

The classical momentum is $\pm$ p and so $\langle (p - \langle p\rangle)^2 \rangle = \langle p^2 \rangle = p^2 = 2mE$

(ii) For the quantum mechanical solution, the energy is high when n is large. In this case $\langle (z - \langle z\rangle)^2\rangle$ tends to the classical solution. For $\langle (p - \langle p\rangle)^2\rangle$, the classical and quantum mechanical solutions agree for all values of n. The reason is that we are dealing with an energy eigenstate, and p^2 is an energy eigenvalue (it accounts for the whole of the energy).

3.4 $$J(x) = -\frac{i\hbar e}{2m}\left(\Psi^* \frac{\partial \Psi}{\partial x} - \Psi \frac{\partial \Psi^*}{\partial x}\right)$$

$$\nabla \cdot \mathbf{J} = \frac{\partial J}{\partial x} \quad \text{(in one dimension)}$$

$$= -\frac{i\hbar e}{2m}\frac{\partial}{\partial x}\left(\Psi^* \frac{\partial \Psi}{\partial x} - \Psi \frac{\partial \Psi^*}{\partial x}\right)$$

$$= -\frac{i\hbar e}{2m}\left(\frac{\partial \Psi^*}{\partial x}\frac{\partial \Psi}{\partial x} + \Psi^* \frac{\partial^2 \Psi}{\partial x^2} - \frac{\partial \Psi}{\partial x}\frac{\partial \Psi^*}{\partial x} - \Psi \frac{\partial^2 \Psi^*}{\partial x^2}\right)$$

$$= -\frac{i\hbar e}{2m}\left(\Psi^* \frac{\partial^2 \Psi}{\partial x^2} - \Psi \frac{\partial^2 \Psi^*}{\partial x^2}\right) \tag{1}$$

From the Schrödinger equation : $\dfrac{\partial^2 \Psi}{\partial x^2} = -\dfrac{2m}{\hbar^2}\left(i\hbar \dfrac{\partial \Psi}{\partial t} - V\Psi\right)$

$$\therefore \Psi^* \frac{\partial^2 \Psi}{\partial x^2} = -\frac{2m}{\hbar^2}\left(i\hbar\Psi^* \frac{\partial \Psi}{\partial t} - V\,|\Psi|^2\right)$$

Subtract the corresponding complex conjugates on both sides,

$$\therefore \Psi^* \frac{\partial^2 \Psi}{\partial x^2} - \Psi \frac{\partial^2 \Psi^*}{\partial x^2} = -\frac{2m}{\hbar^2}\left(i\hbar\Psi^* \frac{\partial \Psi}{\partial t} - V\,|\Psi|^2 + i\hbar\Psi \frac{\partial \Psi^*}{\partial t} + V\,|\Psi|^2 \right)$$

$$= -\frac{2mi}{\hbar}\left(\Psi^* \frac{\partial \Psi}{\partial t} + \Psi \frac{\partial \Psi^*}{\partial t} \right)$$

$$= -\frac{2mi}{\hbar} \frac{\partial}{\partial t}|\Psi|^2$$

Substitute into eqn (1) to get

$$\frac{\partial J}{\partial x} = -e \frac{\partial}{\partial t}|\Psi|^2$$

But by definition, charge density $\rho = e|\Psi|^2$, leading to

$$\frac{\partial J}{\partial x} = -\frac{\partial \rho}{\partial t} = -e\frac{\partial N}{\partial t} \qquad \text{(Q.E.D.)}$$

3.5 In region 1, from [eqns 3.21, 3.22],

$$\psi_1 = A \exp ik_1x + B\exp(-ik_1x) \qquad \text{where } k_1^2 = \frac{2mE}{\hbar^2} \tag{1}$$

In region 2, from [eqns 3.24, 3.25],

$$\psi_2 = C \exp ik_2x \qquad \text{where } k_2^2 = \frac{2m}{\hbar^2}(E - V_2) \tag{2}$$

$$\text{Now } \quad J(x) = -\frac{i\hbar e}{2m}\left(\Psi^* \frac{\partial \Psi}{\partial x} - \Psi \frac{\partial \Psi^*}{\partial x} \right) \tag{3}$$

With eqns (1) and (2),

$$J_1 = \frac{\hbar e}{m}\left(k_1|A|^2 - k_1|B|^2 \right) \quad \text{and}$$

$$J_2 = -\frac{i\hbar e}{2m}\left(ik_2|C|^2 \exp i(k_2 - k_2^*)x + ik_2^*|C|^2 \exp i(k_2 - k_2^*)x \right)$$

$$= \frac{\hbar e}{m} k_{2r}|C|^2 \exp(-2k_{2i}x)$$

where k_{2r}, k_{2i} are the real and imaginary parts of k_2.

A, B, and C are related by [eqn 3.28] (obtained by matching the boundary conditions : $\psi_1 = \psi_2$ and $\frac{\partial\psi_1}{\partial x} = \frac{\partial\psi_2}{\partial x}$ at $x = 0$) :

$$\frac{B}{A} = \frac{k_1 - k_2}{k_1 + k_2} \quad \text{and} \quad \frac{C}{A} = \frac{2k_1}{k_1 + k_2}$$

Here, A represents the amplitude of the incident wave, B that of the reflected wave, and C that of the transmitted wave.

Case (i) $E < V_2$

In this case, k_2 is purely imaginary, so $J_2 = 0$ (because $k_{2r} = 0$). In other words, the transmitted current = 0.

$$\frac{\text{reflected current}}{\text{incident current}} = \frac{J_{1B}}{J_{1A}} = \left|\frac{B}{A}\right|^2 = \left|\frac{k_1 - k_2}{k_1 + k_2}\right|^2 = 1 \ (\because k_1 \text{ is real})$$

Case (ii) $E > V_2$

k_2 is real in this case and

$$\frac{\text{reflected current}}{\text{incident current}} = \left|\frac{B}{A}\right|^2 = \left|\frac{k_1 - k_2}{k_1 + k_2}\right|^2$$

$$\frac{\text{transmitted current}}{\text{incident current}} = \frac{J_2}{J_{1A}} = \frac{k_{2r}|C|^2}{k_1\ |A|^2} = \frac{k_{2r}}{k_1}\left|\frac{2k_1}{k_1 + k_2}\right|^2$$

These equations are the same as that in Ex 1.6.

3.6 Let the solutions in the three regions be

$$\begin{cases} \psi_1 = A \exp ik_1x + B \exp(-ik_1x) \\ \psi_2 = C \exp ik_2x + D \exp(-ik_2x) \\ \psi_3 = F \exp ik_1(x - d) \qquad \text{(note that } k_3 = k_1) \end{cases}$$

where $k_1^2 = \dfrac{2mE}{\hbar^2}$; $k_2^2 = \dfrac{2m}{\hbar^2}(E - V_2)$

From the boundary conditions :

(i) at $x = 0$, $\psi_1 = \psi_2$ and $\dfrac{\partial\psi_1}{\partial x} = \dfrac{\partial\psi_2}{\partial x}$

(ii) at $x = d$, $\psi_2 = \psi_3$ and $\dfrac{\partial\psi_2}{\partial x} = \dfrac{\partial\psi_3}{\partial x}$

we get

$$A + B = C + D \tag{1}$$

$$k_1(A - B) = k_2(C - D) \tag{2}$$

$$C \exp ik_2d + D \exp(-ik_2d) = F \tag{3}$$

$$k_2\left(C \exp ik_2d - D \exp(-ik_2d)\right) = k_1F \tag{4}$$

Our aim is to find the ratio of transmitted current to incident current, which is given by $\left|\dfrac{F}{A}\right|^2$.

Eliminate B from eqns (1) and (2),

$$2A = (1 + k_2/k_1)\, C + (1 - k_2/k_1)\, D \tag{5}$$

From eqns (3) and (4),

$$2C \exp ik_2d = (1 + k_1/k_2)\, F \tag{6}$$

$$2D \exp(-ik_2d) = (1 - k_1/k_2)\, F \tag{7}$$

Substitute C and D from eqns (6) and (7) into eqn (5),

$$4A/F = (1 + k_2/k_1)(1 + k_1/k_2) \exp(-ik_2d)$$

$$+ (1 - k_2/k_1)(1 - k_1/k_2) \exp ik\ d$$

$$= 2 \left(\exp ik_2 d + \exp(-ik_2 d) \right)$$

$$- (k_2/k_1 + k_1/k_2)\cdot \left(\exp ik_2 d - \exp(-ik_2 d) \right)$$

$$= 4 \cosh k_{2i} d + 2i\ (k_{2i}/k_1 - k_1/k_{2i}) \sinh k_{2i} d$$

where $k_{2i} = ik_2$.

Note that since $V_2 > E$, k_2 is purely imaginary or k_{2i} is real.

$$\text{Thus,}\ \frac{J_3}{J_1} = \left|\frac{F}{A}\right|^2 = \left[\cosh^2(k_{2i}d) + \frac{1}{4}\,(k_{2i}/k_1 - k_1/k_{2i})^2 \sinh^2(k_{2i}d) \right]^{-1}$$

For $V_2 = 2.5$ eV, $E = 0.5$ eV,

$J_3/J_1 = \underline{0.136}$ when $d = 2$ Å, and $\underline{6.58\times10^{-13}}$ when $d = 20$ Å

3.7 Put in the numerical values of V_2 and E given in Ex 3.6,

$\exp(-2k_{2i}d) = 0.055$ and 2.57×10^{-13} for $d = 2$ and 20 Å respectively.

These values are of the same orders as that calculated exactly in Ex 3.6. Therefore we can conclude that the approximation is not bad.

3.8 Since the electron energy (1 eV) is smaller than V_3 (3 eV), k_3 is purely imaginary which means that the beam cannot pass through the barrier. Thus all the power is bound to be reflected despite the fact that there is a potential trough in between (this affects the phase of the reflected beam). Hence the magnitude of the reflection coefficient must be unity.

3.9 The time independent Schrödinger equation is

$$\frac{\hbar^2}{2m}\nabla^2\psi + (E - V)\psi = 0 \tag{1}$$

For a two-dimensional rigid potential well

$$V = \begin{cases} 0 & \text{when } |x| < L_x/2 \text{ and } |y| < L_y/2 \\ \infty & \text{otherwise} \end{cases}$$

Let $\psi = \psi_x(x)\psi_y(y)$ where ψ_x is a function of x only, ψ_y is a function of y only. Substitute into eqn (1) and rearrange,

$$\frac{\hbar^2}{2m\psi_x}\frac{\partial^2\psi_x}{\partial x^2} + E = -\frac{\hbar^2}{2m\psi_y}\frac{\partial^2\psi_y}{\partial y^2} \quad \text{for } |x| < L_x/2 \text{ and } |y| < L_y/2$$

Since L.S. is a function of x only while R.S. is a function of y only, they must be equal to a constant, E_2 say.

$$\frac{\hbar^2}{2m}\frac{\partial^2\psi_x}{\partial x^2} + E_1\psi_x = 0 \quad \text{and} \quad \frac{\hbar^2}{2m}\frac{\partial^2\psi_y}{\partial y^2} + E_1\psi_y = 0 \quad \text{where } E = E_1 + E_2$$

These two equations are of the same form as the Schrödinger equation. Let the solutions be

$$\begin{cases} \psi_x = A\cos k_x x + B\sin k_x x \\ \psi_y = C\cos k_y y + D\sin k_y y \end{cases}$$

where $k_x = \sqrt{2mE_1}/\hbar$, $k_y = \sqrt{2mE_2}/\hbar$; A, B, C, D are constants.

At $x = \pm L_x/2$, $\psi = 0$, and at $y = \pm L_y/2$, $\psi = 0$,

$\therefore$ we can have two cases for ψ_x :

(i) $A = 0$ and $k_x L_x/2 = n\pi$ where n is an integer (anti-symmetric solution)

(ii) $B = 0$ and $k_x L_x/2 = (2m + 1)\pi/2$ where m is an integer (symmetric solution)

Summarizing the two cases, $k_x L_x = n\pi$ where n is an integer. Similarly $k_y L_y = m\pi$ where m is an integer.

$$\text{Hence} \quad E = E_1 + E_2 = \frac{\hbar^2}{2m}\left(k_x^2 + k_y^2\right) = \frac{\hbar^2\pi^2}{2m}\left(\frac{n^2}{L_x^2} + \frac{m^2}{L_y^2}\right)$$

When $L_y = \frac{3}{2} L_x$, the energy of the first five lowest states are $4E_o/9$, E_o, $13E_o/9$, $25E_o/9$, and $4E_o$ where $E_o = \dfrac{\hbar^2\pi^2}{2mL_x^2}$

3.10 The anti-symmetric solution is

$$\begin{cases} \psi_1 = -C \exp \gamma z \\ \psi_2 = A \sin kz \\ \psi_3 = C \exp(-\gamma z) \end{cases}$$

where $\gamma^2 = \dfrac{2m}{\hbar^2}(V_1 - E)$, $k^2 = \dfrac{2m}{\hbar^2} E$

At $z = -L/2$, $\psi_1 = \psi_2$ and $\dfrac{\partial\psi_1}{\partial z} = \dfrac{\partial\psi_2}{\partial z}$

$$\therefore \begin{cases} A \sin kL/2 = C \exp(-\gamma L/2) \\ -Ak \cos kL/2 = C\gamma \exp(-\gamma L/2) \end{cases}$$

For non-trivial solutions for A and C,

$$k \cot kL/2 + \gamma = 0$$

$$\text{or} \quad \sqrt{E} + \sqrt{V - E} \tan\left(\frac{mL^2}{2\hbar^2} E\right)^{1/2} = 0 \qquad (1)$$

The energy of the lowest state for $L = 10^{-9}$ m, $V_1 = 1.6\times10^{-18}$ J can be obtained by solving eqn (1) graphically like [Fig 3.5], or by trial and error as we did here (note that $\left(\frac{mL^2}{2\hbar^2}E\right)^{1/2}$ is between $\pi/2$ and π for the lowest state), giving $E \simeq \underline{1.9\times10^{-19}\ J}$

3.11 Let the wavefunctions be

$$\begin{cases} \psi_1 = A\cos k_1x \\ \psi_2 = B\cos k_2x + C\sin k_2x \end{cases}$$

where $k_1^2 = 2m(E - V_0)/\hbar^2$ and $k_2^2 = 2mE/\hbar^2$

(ψ_2 is the wavefunction in the region $b < x < a$. Since only symmetric functions are considered, this means that in the region $-a < x < b$ the wavefunction is $B\cos k_2x - C\sin k_2x$, but that need not concern us, we shall do the calculations only for positive x)

Note that depending on the magnitude of E, k_1 is not necessarily real.

At $x = a$, $\psi_2 = 0$,

$$\therefore\ B\cos k_2a + C\sin k_2a = 0$$

At $x = b$, $\psi_1 = \psi_2$ and $\frac{\partial\psi_1}{\partial x} = \frac{\partial\psi_2}{\partial x}$

$$\therefore \begin{cases} A\cos k_1b = B\cos k_2b + C\sin k_2b \\ -k_1A\sin k_1b = -k_2B\sin k_2b + k_2C\cos k_2b \end{cases}$$

In order to have non-trivial solutions for A, B and C

$$\begin{vmatrix} 0 & \cos k_2a & \sin k_2a \\ \cos k_1b & -\cos k_2b & -\sin k_2b \\ k_1\sin k_1b & -k_2\sin k_2b & k_2\cos k_2b \end{vmatrix} = 0$$

The roots of it represent the energy states that can exist in the potential well given.

The determinant can in fact be expanded and simplified to

$$(k_1/k_2)\tan k_1 b \tan k_2(a-b) = 0$$

3.12 With the assumption of exp(-iωt) time dependence and $\sigma = 0$ (lossless case), [eqn 1.27] and [eqn 1.29] reduce to

$$\frac{\partial \mathscr{E}_x}{\partial z} = i\omega B_y \quad \text{and} \quad -\frac{\partial B_y}{\partial z} = -i\omega\mu\varepsilon\mathscr{E}_x$$

Eliminate B_y to get $\dfrac{\partial^2 \mathscr{E}_x}{\partial z^2} + \omega^2\mu\varepsilon\mathscr{E}_x = 0$ which, apart from the constants, has the same form as the time-independent Schrödinger equation for constant potential [eqn 3.17]. (Q.E.D.)

The Poynting vector in electromagnetic theory is defined as

$\mathbf{P_d} = \frac{1}{2}\,\mathrm{Re}\{\boldsymbol{\mathscr{E}} \times \mathbf{H}^*\}$ which for the plane wave $(\mathscr{E}_x, H_y)$ reduces to

$$\mathbf{P_d} = \frac{1}{2}\,\mathrm{Re}\left\{\mathscr{E}_x\cdot\left(\frac{1}{i\omega\mu}\cdot\frac{\partial \mathscr{E}_x}{\partial z}\right)^*\right\} \quad \text{in the z direction}$$

$$= \frac{1}{4}\left\{\mathscr{E}_x\cdot\left(\frac{1}{i\omega\mu}\cdot\frac{\partial \mathscr{E}_x}{\partial z}\right)^* + \text{complex conjugate}\right\}$$

$$= \frac{1}{4}\left\{-\frac{1}{i\omega\mu}\,\mathscr{E}_x\,\frac{\partial \mathscr{E}_x^*}{\partial z} + \frac{1}{i\omega\mu}\,\mathscr{E}_x^*\,\frac{\partial \mathscr{E}_x}{\partial z}\right\}$$

$$= -\frac{i}{4\omega\mu}\left(\mathscr{E}_x^*\,\frac{\partial \mathscr{E}_x}{\partial z} - \mathscr{E}_x\,\frac{\partial \mathscr{E}_x^*}{\partial z}\right)$$

which is of the same functional form as that for the quantum mechanical current defined in Ex 3.4.

3.13 Introduce new variable $\zeta = \alpha x$, the time independent Schrödinger equation becomes $\frac{\hbar^2\alpha^2}{2m}\frac{\partial^2\psi}{\partial\zeta^2} + (E - V)\psi = 0$

With $\psi = H_n \exp(-\frac{1}{2}\zeta^2)$, $V = \frac{1}{2} m\omega_o^2 x^2$, $\alpha^2 = \frac{m\omega_o}{\hbar}$

$$E = V - \frac{\hbar^2\alpha^2}{2m}\cdot\frac{H_n'' - 2\zeta H_n' + (\zeta^2 - 1)H_n}{H_n}$$

$$= \frac{1}{2}\hbar\omega_o\zeta^2 - \frac{1}{2}\hbar\omega_o\,\frac{H_n'' - 2\zeta H_n' + (\zeta^2 - 1)H_n}{H_n}$$

$$= \frac{1}{2}\hbar\omega_o\,\frac{-H_n'' + 2\zeta H_n' + H_n}{H_n}$$

Put in the expressions of H_n yields

$$E_o = \frac{1}{2}\hbar\omega_o , \quad E_1 = \frac{3}{2}\hbar\omega_o , \quad E_2 = \frac{5}{2}\hbar\omega_o , \quad E_3 = \frac{7}{2}\hbar\omega_o$$

or $E_n = (n + \frac{1}{2})\hbar\omega_o$

It agrees with Planck's formula for the energy of photons but for the so-called 'zero point energy' at n = 0.

4. The hydrogen atom and the periodic table

4.1 The energy difference between 1s and 2s states is given by

$$\Delta E = E_2 - E_1 = 13.6\ (1 - 1/2^2)\ \text{eV} \quad \text{(using [eqn 4.26])}$$

The required frequency of electromagnetic wave is given by $hf = \Delta E$

$$\therefore\ \lambda = hc/\Delta E = \frac{6.62\text{x}10^{-34}\text{x}3\text{x}10^{8}}{13.6\text{x}(1 - \frac{1}{4})\text{x}1.6\text{x}10^{-19}} = \underline{1.22\text{x}10^{-7}\ \text{m}}$$

4.2 By the Conservation of Energy,

incident photon energy = ionization energy + kinetic energy of the ionized electron

Thus $hc/\lambda = 13.6\ \text{eV} + \frac{1}{2}mv^2$ where v is the maximum speed at which an ionized electron may leave the atom

$$\therefore\ \frac{6.62\text{x}10^{-34}\text{x}3\text{x}10^{8}}{20\text{x}10^{-9}} = 13.6\text{x}1.6\text{x}10^{-19} + \tfrac{1}{2}\text{x}9.11\text{x}10^{-31}\ v^2$$

$$v = \underline{4.13\text{x}10^{6}\ \text{ms}^{-1}}$$

4.3 If the lifetime of the state was infinitely long then the transition would be infinitely sharp with a zero bandwidth. But a finite lifetime implies oscillation at the transition frequency for a finite time which means a finite bandwidth (cf. Ex 2.2).

Using the uncertainty relationship the argument is that if the uncertainty, whether the electron is in the upper or in the lower state, is equal to the lifetime then the corresponding energy width of the transition is $\Delta E = h/\Delta t$ and the uncertainty in the wavelength (i.e. the bandwidth) is given by

$$\Delta(hc/\lambda) = h/\Delta t \quad \text{or} \quad -c\ \Delta\lambda/\lambda^2 = 1/\Delta t$$

$$\therefore\ \Delta\lambda = -\lambda^2/(c\Delta t) = -(450\times10^{-9})^2/(3\times10^8\times10^{-8})\ \text{m} = \underline{6.75\times10^{-5}\ \text{nm}}$$

4.4 The total energy of the electron, E_{total} = kinetic energy + potential energy. The potential energy is due to electrostatic interaction and is equal to $-\dfrac{e^2}{4\pi\varepsilon_0 r_m}$.

$$\therefore\ E_{total} = \frac{\hbar^2}{2mr_m^2} - \frac{e^2}{4\pi\varepsilon_0 r_m}$$

In order that it is minimum, $\dfrac{\partial}{\partial r_m}E_{total} = 0$

$$\text{i.e.}\quad -\frac{\hbar^2}{mr_m^3} + \frac{e^2}{4\pi\varepsilon_o r_m^2} = 0$$

$$r_m = \frac{4\pi\hbar^2\varepsilon_o}{me^2}$$, the same as the first Bohr radius given in [eqn 4.24].

Note that we have quoted here the uncertainty principle in the form $\Delta p\Delta z \lesssim \hbar$ in order to get the right formula for the Bohr radius whereas formerly $\Delta p\Delta z \lesssim h$ was used. The point is that the elementary derivations which do not employ a rigorous definition of the term "uncertainty" cannot be expected to yield an exact value for the right-hand-side.

4.5 This is actually an application of [eqn 3.43]. In the present case r is not an operator, therefore the formula is reduced to

$$\langle r \rangle = \int r|\psi|^2\, dV \Big/ \int |\psi|^2\, dV$$

$$= \int_0^\infty r^3 \exp(-2C_o r)\, dr \Big/ \int_0^\infty r^2 \exp(-2C_o r)\, dr$$

$$= 3/2C_o = \underline{0.0792\ \text{nm}}\ \text{(from [eqn 4.24])}$$

4.6 (i) It means that the quantum numbers are $n = 2$, $\ell = 0$, $m_\ell = 0$. For spherically symmetric solutions (no dependence on θ and ϕ) both ℓ and m_ℓ must be zero.

(ii), (iii) Substitute [eqn 4.32] into the time-independent Schrödinger equation for the spherically symmetric case [eqn 4.12],

$$\frac{\hbar^2}{2m}\left\{ -C_oC_1 + \frac{C_o^2}{4}(1 + C_1 r) + \frac{2}{r}\left[C_1 - \frac{C_o}{2}(1 + C_1 r) \right] \right\} A \exp(-C_o r/2)$$

$$+\left(E + \frac{e^2}{4\pi\varepsilon_o r} \right)(1 + C_1 r) A \exp(-C_o r/2) = 0$$

Since it is valid for all r, the coeffs. of $r \exp(-C_o r/2)$, $\exp(-C_o r/2)$ and $\exp(-C_o r/2)/r$ must vanish. Therefore

$$\begin{cases} \dfrac{\hbar^2}{2m}\dfrac{C_o^2 C_1}{4} + EC_1 = 0 \\[2ex] \dfrac{\hbar^2}{2m}(C_o^2/4 - 2C_oC_1) + E + \dfrac{e^2}{4\pi\varepsilon_o} = 0 \\[2ex] \dfrac{\hbar^2}{2m}(2C_1 - C_o) + \dfrac{e^2}{4\pi\varepsilon_o} = 0 \end{cases}$$

giving (for $C_1 \neq 0$)

$$C_o = \frac{me^2}{4\pi\hbar^2\varepsilon_o}, \quad C_1 = -\frac{me^2}{8\pi\hbar^2\varepsilon_o} \quad \text{and} \quad E = -\frac{me^4}{32\varepsilon_o^2 h^2}$$

(iv) The probability that the electron can be found in the spherical shell between r and r + dr is $4\pi r^2|\psi|^2 dr$. Since the total probability of finding the electron somewhere must be unity,

$$1 = \int_0^\infty 4\pi r^2|\psi|^2\, dr = 4\pi A^2 \int_0^\infty r^2(1 + C_1 r)^2 \exp(-C_o r)\, dr$$

$$= 8\pi A^2/C_o^3 \quad \text{(note that } C_1 = -C_o/2)$$

$$\therefore \quad A = \left(C_o^3/8\pi\right)^{1/2}$$

(v) The probability distribution is proportional to

$$r^2(1 + C_1 r)^2 \exp(-C_o r)$$

The slope is zero when $2(r + C_1 r^2)(1 + 2C_1 r) - C_o(r + C_1 r^2)^2 = 0$

Solving in terms of C_o (note that $C_1 = -C_o/2$) to get

$$r = 0,\ 2/C_o \text{ or } (3 \pm \sqrt{5})/C_o$$

By accessing the change of slope in the vicinity of these values of r, it can be shown that there are minima at $r = 0$ and $2/C_o$ (the function values are both zero), and there are maxima at $r = (3 \pm \sqrt{5})/C_o$. The value at $r = (3 + \sqrt{5})/C_o$ is larger than that of $r = (3 - \sqrt{5})/C_o$. Therefore the most probable orbit of the electron is at $r = (3 + \sqrt{5})/C_o$. The above consideration is consistent with the curve given in [Fig 4.3] for n = 2.

4.7 In spherical coordinates (r, θ, ϕ),

$$\nabla^2\psi = \frac{1}{r^2}\frac{\partial}{\partial r}\left(r^2\frac{\partial\psi}{\partial r}\right) + \frac{1}{r^2\sin\theta}\frac{\partial}{\partial\theta}\left(\sin\theta\frac{\partial\psi}{\partial\theta}\right) + \frac{1}{r^2\sin^2\theta}\frac{\partial^2\psi}{\partial\phi^2}$$

For $\psi = \psi_{210} = r \exp(-C_o r/2) \cos\theta$,

$$\nabla^2\psi = \frac{\cos\theta}{r^2}\frac{\partial}{\partial r}\left(r^2(1 - C_o r/2)\exp(-C_o r/2) \right) + \frac{\exp(-C_o r/2)}{r \sin\theta}\frac{\partial}{\partial\theta}(-\sin^2\theta)$$

$$= (C_o^2/4 - 2C_o/r)\, r \exp(-C_o r/2)\cos\theta$$

$$\therefore \frac{\hbar^2}{2m}\nabla^2\psi + \left(E + \frac{e^2}{4\pi\varepsilon_o r} \right)\psi = \left[\frac{\hbar^2}{2m}\left(C_o^2/4 - 2C_o/r \right) + \left(E + \frac{e^2}{4\pi\varepsilon_o r} \right) \right]\psi$$

$$= 0 \text{ when } C_o = \frac{me^2}{4\pi\hbar^2\varepsilon_o} \text{ and } E = -\hbar^2 C_o^2/8m$$

Therefore the Schrödinger equation is satisfied and the corresponding energy $E = -\hbar^2 C_o^2/8m = -me^4/(32\varepsilon_o^2 h^2)$. This energy is the same as that of ψ_{200} wavefunction.

4.8 The Schrödinger equation for helium (two protons and two electrons) is given by [eqn 4.30] :

$$-\frac{\hbar^2}{2m}(\nabla_1^2\psi + \nabla_2^2\psi) + \frac{1}{4\pi\varepsilon_o}\left(-\frac{2e^2}{r_1} - \frac{2e^2}{r_2} + \frac{e^2}{r_{12}} \right)\psi = E\psi \qquad (1)$$

Neglect the potential term between the two electrons (i.e. $\frac{e^2}{4\pi\varepsilon_o r_{12}}$) and let $\psi(r_1, r_2) = \psi_1(r_1)\,\psi_2(r_2)$.

Eqn (1) can be rearranged to get

$$\left(-\frac{\hbar^2}{2m}\frac{\nabla_1^2\psi_1}{\psi_1} - \frac{1}{4\pi\varepsilon_o}\frac{2e^2}{r_1} \right) + \left(-\frac{\hbar^2}{2m}\frac{\nabla_2^2\psi_2}{\psi_2} - \frac{1}{4\pi\varepsilon_o}\frac{2e^2}{r_2} \right) = E$$

The terms in the first bracket are functions of r_1 only while the terms in the second bracket are functions of r_2 only. Therefore,

$$-\frac{\hbar^2}{2m}\frac{\nabla_i^2\psi_i}{\psi_i} - \frac{1}{4\pi\varepsilon_o}\frac{2e^2}{r_i} = E_i \quad (i = 1,\ 2), \text{ and } \quad E_1 + E_2 = E$$

where E_1, E_2 are constants.

These equations are in fact identical to the Schrödinger equation of hydrogen atom when e^2 is replaced by $2e^2$.

Hence $E_1 = E_2 = -\dfrac{m(2e^2)^2}{8\varepsilon_o^2h^2}$ (from [eqn 4.18]) $= -\ 4\text{x}13.6$ eV = $\underline{-54.4 \text{ eV}}$

This energy is very much below the measured -24.6 eV. This is because we have neglected the interaction between the electrons. If it is taken into account, the electrons are less tightly bounded because of their mutual repulsion, and thus the magnitude of the energy will be smaller.

4.9 A lithium atom has three protons and three electrons. So ∇^2 should be replaced by $(\nabla_1^2 + \nabla_2^2 + \nabla_3^2)$. The potential energy of electron i at a distance r_i from the protons is $-\dfrac{3e^2}{4\pi\varepsilon_o r_i}$ and that due to electron j is $\dfrac{e^2}{4\pi\varepsilon_o r_{ij}}$ where r_{ij} is the distance between electron i and electron j.

Therefore the total potential energy $= \sum_{i=1}^{3} -\dfrac{3e^2}{4\pi\varepsilon_o r_i} + \sum_{i\neq j}\dfrac{e^2}{4\pi\varepsilon_o r_{ij}}$.

Hence the corresponding time independent Schrödinger equation is

$$-\frac{\hbar^2}{2m}(\nabla_1^2 + \nabla_2^2 + \nabla_3^2)\psi +$$

$$\frac{1}{4\pi\varepsilon_o}\left[-\frac{3e^2}{r_1} - \frac{3e^2}{r_2} - \frac{3e^2}{r_3} + \frac{e^2}{r_{12}} + \frac{e^2}{r_{13}} + \frac{e^2}{r_{23}}\right]\psi = E\psi$$

5. Bonds

5.2 Force on the positive centre of one dipole due to the other dipole

$$= \frac{q^2}{4\pi\varepsilon_o}\left(\frac{1}{r^2} - \frac{1}{(r + d)^2}\right) \quad \text{where } q = \text{charge of the dipole}$$

Force on the negative centre $= \frac{q^2}{4\pi\varepsilon_o}\left(\frac{1}{r^2} - \frac{1}{(r - d)^2}\right)$

Thus the total force on the dipole

$$= \frac{q^2}{4\pi\varepsilon_o}\left(\frac{1}{r^2} - \frac{1}{(r + d)^2} + \frac{1}{r^2} - \frac{1}{(r - d)^2}\right)$$

$$= \frac{q^2}{4\pi\varepsilon_o r^2}\left(2 - (1 + d/r)^{-2} - (1 - d/r)^{-2}\right)$$

$$\simeq \frac{q^2}{4\pi\varepsilon_o r^2}\left[2 - (1 - 2d/r + 3d^2/r^2) - (1 + 2d/r + 3d^2/r^2)\right] \quad (\because d << r)$$

$$= -\frac{3q^2d^2}{2\pi\varepsilon_o r^4}$$

Thus the force varies as r^{-4}.

5.3 $E(r) = A/r^n - B/r^m$ [eqn 5.8]

The atoms are in equilibrium when the energy is a minimum,

$$\text{i.e. } \frac{\partial E}{\partial r} = -nAr^{-n-1} + mBr^{-m-1} = 0$$

$\therefore$ the equilibrium distance r_o is given by $r_o^{n-m} = \frac{nA}{mB}$ (1)

The atoms can be pulled apart when the applied force is larger than the maximum interatomic force. Note that force $= -$ grad $E = -\frac{\partial E}{\partial r}$ (i.e. the slope of the E(r) curve). Hence the interatomic force is maximum when $\frac{\partial^2 E}{\partial r^2}$ = 0 (in other words, it is at the point of inflexion of E(r) curve).

i.e. $n(n + 1)Ar^{-(n+2)} - m(m + 1)Br^{-(n+2)} = 0$

giving $r^{n-m} = \frac{n(n + 1)A}{m(m + 1)B} = \frac{n + 1}{m + 1} r_o^{n-m}$ (from eqn (1))

Hence $r_b = \left(\frac{n + 1}{m + 1} \right)^{1/(n-m)} r_o$ (Q.E.D.)

5.4 From [eqn 5.7], the bulk modulus of elasticity $c = \frac{1}{9r_o} \frac{\partial^2 E}{\partial r^2} \bigg|_{r=r_o}$

With $E(r) = A/r^n - B/r^m$,

$$c = \left(n(n + 1)Ar_o^{-n-2} - m(m + 1)Br_o^{-m-2} \right) / 9r_o$$

$$= \frac{m(n - m)B}{9r_o^{m+3}} \quad \text{(from eqn (1) of Ex 5.3)} \qquad (1)$$

Hence the separation of K^+Cl^- ions,

$$r_o = \left(\frac{m(n - m)B}{9c} \right)^{1/(m+3)}$$

$$= \left(\frac{1x8x1.75x(1.6x10^{-19})^2 \ / \ 2x4\pi x8.85x10^{-12}}{9x1.88x10^{10}} \right)^{1/4} = \underline{3.12x10^{-10}\ m}$$

5.5 From [eqns 5.7 and 5.9],

$$c = \frac{1}{9r_o} \frac{\partial^2}{\partial r^2} \left(A/r^n - B/r^m \right) \text{ at } r = r_o$$

$$= \frac{1}{9r_o} \left(An(n + 1)/r_o^{n+2} - Bm(m + 1)/r_o^{m+2} \right)$$

At $r = r_o$, $\frac{\partial E}{\partial r} = 0$

$\therefore \quad - An/r_o^{n+1} + Bm/r_o^{m+1} = 0$

Hence $c = \frac{1}{9r_o}\left(Bm(n+1)/r_o^{m+2} - Bm(m+1)/r_o^{m+2} \right)$

$$= - \frac{mn}{9r_o^3}\frac{B}{r_o^m}\left(\frac{m}{n} - 1\right) = - \frac{mn}{9r_o^3} E_c \quad \text{(from [eqn 5.9])} \qquad \text{(Q.E.D.)}$$

5.6 The energy between two ions at a distance r is $\pm \frac{q_1 q_2}{4\pi\varepsilon_o |r|}$ where q_1, q_2 are the charges of the two ions. Now there are charges of opposite sign at $r = \pm r_o, \pm 3r_o, \pm 5r_o, \cdots$ and equal sign at $r = \pm 2r_o, \pm 4r_o, \pm 6r_o, \cdots$ where r_o is the distance between the ions.

$\therefore$ total energy $= \frac{2q^2}{4\pi\varepsilon_o}\left(- \frac{1}{r_o} + \frac{1}{2r_o} - \frac{1}{3r_o} + \cdots \right) = \underline{- \frac{q^2}{2\pi\varepsilon_o}\log 2}$

5.7 The differential equations describing the system are given by [eqns 5.25 and 5.26] :

$$\begin{cases} i\hbar \frac{d\omega_1}{dt} = H_{11}\omega_1 + H_{12}\omega_2 \\ i\hbar \frac{d\omega_2}{dt} = H_{21}\omega_1 + H_{22}\omega_2 \end{cases}$$

Let the solution be $\omega_n = K_n \exp(-iEt/\hbar)$ where $n = 1, 2$ [eqn 5.33]

Substitute into the differential equations,

$$\begin{cases} K_1 E = H_{11}K_1 + H_{12}K_2 \\ K_2 E = H_{21}K_1 + H_{22}K_2 \end{cases}$$

which has non-trivial solution only if

$$\begin{vmatrix} E - H_{11} & -H_{12} \\ -H_{21} & E - H_{22} \end{vmatrix} = 0$$

yielding $E_{1,2} = \dfrac{H_{11} + H_{22}}{2} \mp \left\{ \left(\dfrac{H_{22} - H_{11}}{2} \right)^2 + A^2 \right\}^{1/2}$ $\quad (\because H_{12}H_{21} = A^2)$

If $A = 0$ (without coupling), $E_1 = H_{11}$ and $E_2 = H_{22}$.

Without loss of generality, assume $H_{11} < H_{22}$. The energy reduced below H_{11} owing to the coupling is

$$H_{11} - \frac{H_{11} + H_{22}}{2} + \left\{ \left(\frac{H_{22} - H_{11}}{2} \right)^2 + A^2 \right\}^{1/2}$$

$$= \left(x^2 + A^2 \right)^{1/2} - x \quad \text{where } x = \frac{H_{22} - H_{11}}{2}$$

and this is less than A since $(A + x)^2 > x^2 + A^2$ as x and A are positive.

(Q.E.D.)

6. The free electron theory of metals

6.1 (i) $F(E) = \left[1 + \exp{(E - E_F)/kT} \right]^{-1}$ (1)

For $E - E_F = kT$, $F(E) = (1 + e)^{-1} = \underline{0.270}$

(ii) When $E - E_F = 0.5$ eV and $F(E) = 1\%$,

from eqn (1), $T = \dfrac{E - E_F}{k \log(1/F - 1)} = \dfrac{0.5}{8.62\times10^{-5} \log(1/0.01 - 1)} = \underline{1262.3 \text{ K}}$

6.2 The main steps in the derivation of the Fermi level are :

(i) solve the Schrödinger equation for the three-dimensional potential well;

(ii) at absolute zero, each energy level below E_F is occupied by two electrons (of opposite spins);

(iii) introduce the density of states (as a function of energy E) to simplify the calculation and integrate from zero to E_F to get the total number of electrons;

(iv) equate this to the number of electrons available and hence E_F can be determined.

From [eqn 6.16], $E_F = \dfrac{h^2}{2m}\left(\dfrac{3N}{8\pi}\right)^{2/3}$

For sodium (results of Ex 1.4), $N = 2.5\times10^{28}$ m^{-3}, $m = 8.976\times10^{-31}$ kg,

$\therefore$ $E_F = 5.06\times10^{-19}$ J or $\underline{3.16 \text{ eV}}$

6.3 The energy of the incident light = hc/λ

$$= 6.62\text{x}10^{-34}\text{x}3\text{x}10^{8}/0.2\text{x}10^{-6}\ \text{J} = 6.2\ \text{eV}$$

Since this energy is larger than the work functions of all the metals listed in [Table 6.2], all of them will emit electrons in response to the input light.

6.4 At an energy kT above Fermi level E_F, the density of occupied states = density of states x Fermi function = $C\ \sqrt{E_F + kT}\ /(1 + e)$ where C is a constant given by [eqn 6.10].

For the same density of occupied states at an energy E below the Fermi level,

$$\sqrt{E_F - E}\ /\ \left[\ 1 + \exp(-E/kT)\ \right] = \sqrt{E_F + kT}\ /\ (1 + e) \qquad (1)$$

Since the change of the density of states is much slower than the change in Fermi function for small E, the solution of eqn (1) must be at a large value of E so that the Fermi function becomes fairly constant and the density of states can compensate the difference due to the change in the Fermi function. Hence eqn (1) can be written approximately as

$$\sqrt{E_F - E} \times 1 = \sqrt{E_F}\ /\ (1 + e) \quad \text{assuming that } kT \ll E_F$$

giving $E = \underline{0.928\ E_F}$ (note that the approximation that the Fermi function $\simeq 1$ is justified with such a value of E as $kT \ll E_F$).

6.5 The energy states of electrons for a two-dimensional metal is

$$E = \left(n_x^2 + n_y^2\right)h^2/8mL^2 \qquad \text{(analogous to [eqn 6.2])}$$

The number of states in a circle of radius n is equal to the numerical value of the area divided by 4 (because only positive values of n_x, n_y are permissible) and multiplied by 2 (to allow for two values of spin)

i.e. no. of states $= \frac{\pi n^2}{2} = \frac{\pi}{2} \frac{8mL^2}{h^2} E$

At absolute zero all the states below Fermi level are occupied by electrons.

$\therefore$ the total number of electrons $= 4\pi mL^2 E_F/h^2$

But this is also equal to NL^2.

$\therefore$ $\underline{E_F = Nh^2/4\pi m}$

6.6 From [eqn 6.10], $Z(E) = CE^{1/2}$ where $C = 4\pi L^3(2m)^{3/2}/h^3$

The average energy of free electrons, $\langle E \rangle$, is

$$\frac{1}{NL^3}\int_0^{\infty} F(E)\, Z(E)\, E\, dE \qquad \text{(from [eqn 6.24])}$$

$$= \frac{1}{NL^3}\int_0^{E_F} C\, E^{3/2}\, dE \qquad \text{at absolute zero}$$

$$= \frac{2C}{5NL^3} E_F^{5/2}$$

$$= \frac{2}{5NL^3} \frac{4\pi L^3(2m)^{3/2}}{h^3} \left[\frac{h^2}{2m}\left(\frac{3N}{8\pi}\right)^{2/3}\right]^{3/2} E_F \qquad \left(\because E_F = \frac{h^2}{2m}\left(\frac{3N}{8\pi}\right)^{2/3}\right)$$

$$= 3E_F/5 \qquad \text{(Q.E.D.)}$$

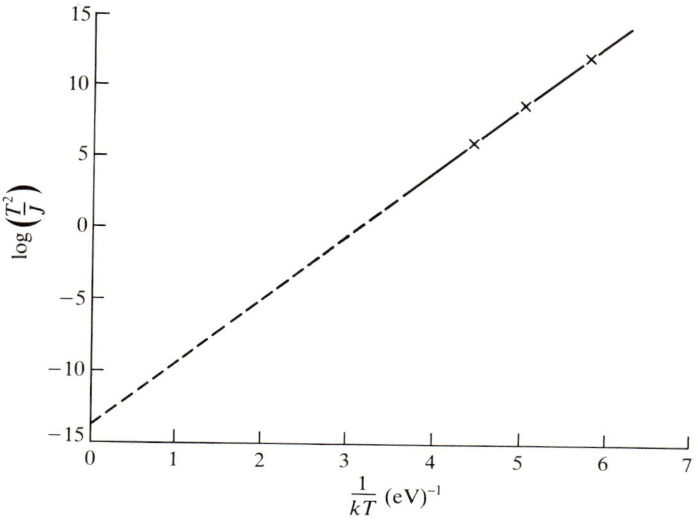

Fig 6.1 Plot of log T^2/J versus $1/kT$.

6.7 (i) Neglect the reflection coefficient in [eqn 6.37],

$$J = A_o T^2 \exp(-\phi/kT)$$

or $\log(T^2/J) = \phi/kT - \log A_o$

Plot $\log(T^2/J)$ versus $1/kT$ (see Fig 6.1). This gives a straight line which shows that the data obey the Richardson law.

The slope of the line = ϕ (the work function) $\approx$ 4.5 eV.

The 'y'-intercept = $-\log A_o = -14.0$,

$\therefore A_o = 1.2\times10^6\ \text{Am}^2\text{K}^{-2}$

Note that as the experimental points lie far away from the 'y'-axis, the intercept cannot be determined accurately. Hence, depending on the chosen 'best-fit' line, the value of A_o obtained may vary appreciably.

(ii) The temperature coefficient α is defined as $\alpha = \frac{1}{R}\frac{dR}{dT}$

Now at 2000 K, R = 3.37/1.60 = 2.106 Ω;

at 2300 K, R = 5.12/1.96 = 2.612 Ω;

at 2600 K, R = 7.40/2.30 = 3.217 Ω.

$$\therefore \text{ mean temp. coeff.} = \frac{3.217 - 2.106}{2.612 \times 600} = \underline{7.1\times10^{-4}\ \mathrm{K}^{-1}}$$

(iii) When the anode voltage is large, it may cause lowering of the effective work function (the Schottky effect). From [eqn 6.45], the thermionic emission will rise by a factor of $\exp\left(e\sqrt{e\mathscr{E}_c/4\pi\varepsilon_o} / kT\right)$ where $\mathscr{E}_c$ is the electric field at the surface of the cathode.

Now $\mathscr{E}_c = \dfrac{V}{r_c \ln(r_a/r_c)}$ where r_c, r_a are the cathode radius and the anode radius respectively; V is the anode voltage (the expression can be derived from Gauss' law : $2\pi r\mathscr{E}(r)$ = constant, and that $\int_{r_a}^{r_c}\mathscr{E}(r)\,dr = V$).

Put in the numerical values,

$$\mathscr{E}_c = \frac{2.3\times10^{3}}{0.0625\times10^{-3}\times\ln(2.5/0.0625)} = 9.98\times10^{6}\ \mathrm{Vm}^{-1}$$

$\therefore$ the emission current will increase by a factor of

$$\exp\left[\left(\frac{1.6\times10^{-19}\times9.98\times10^{6}}{4\pi\times8.85\times10^{-12}}\right)^{1/2} / \ 8.62\times10^{-5}\ T\right]$$

= 2.00 at 2000 K, 1.83 at 2300 K, 1.71 at 2600 K

i.e. it will rise by about <u>90%</u>.

6.8 The number of photons in the laser beam = $\dfrac{\text{laser power}}{\text{energy of a photon}}\ \mathrm{s}^{-1}$

$$= 2\text{x}10^{-3}/\left(\frac{hc}{\lambda}\right) = \frac{2\text{x}10^{-3}\text{x}632.8\text{x}10^{-9}}{6.62\text{x}10^{-34}\text{x}3\text{x}10^{8}} = 6.37\text{x}10^{15}\ \text{s}^{-1}$$

Each photon induces 10^{-4} electrons. Thus the total current induced by the laser beam = $6.37\text{x}10^{15}\text{x}10^{-4}\text{x}1.6\text{x}10^{-19}$ coulomb per second = $\underline{1.02\text{x}10^{-7}\text{ A}}$. Assume that all the electrons come from the Fermi level. Then they come out with a kinetic energy $E_{KE} = hf - \phi$. The electron flow may thus be stopped with an anode voltage so negative that $-eV_a = E_{KE}$. Hence, the work function may be obtained from $\phi = hf - |eV_a|$.

6.9 no. of Cu atoms in 1 m^3, N_a, $= \frac{9.4\text{x}10^3}{63.5}\text{x}6.02\text{x}10^{26} = 8.91\text{x}10^{28}$

Each Cu atom contributes one electron, therefore $N_e = 8.91\text{x}10^{28}\ \text{m}^{-3}$.

$$\text{Hence } E_F = \frac{h^2}{2m}\left(\frac{3N_e}{8\pi}\right)^{2/3} = \underline{7.27\text{ eV}}$$

Electronic specific heat $= \frac{\pi^2 k^2 T}{2E_F} N_e$ / density of Cu (from [eqn 6.25])

$= 2.24\ \text{Jkg}^{-1}\text{K}^{-1}$ (taking room temp. as 293 K)

Lattice specific heat = $3N_a k$ / density of Cu

$= 3.92\text{x}10^2\ \text{Jkg}^{-1}\text{K}^{-1}$

So the total specific heat = $2.24 + 3.92\text{x}10^2 = \underline{394\ \text{Jkg}^{-1}\text{K}^{-1}}$, and $2.24/394 \simeq \underline{0.5\%}$ is contributed by the electrons.

In reference books, they give a value of about 398 $\text{Jkg}^{-1}\text{K}^{-1}$ which is close to our estimation.

6.10 (i) When a voltage U is applied, the density of states becomes

$$\begin{cases} Z_1(E) = C_1(E - E_o - eU)^{1/2} & \text{for } E > E_o + eU \\ Z_r(E) = C_r\, E^{1/2} & \text{for } E > 0 \end{cases}$$

So the number of filled states in the 'left' metal (see [Fig 6.14(b)]) is

$$\begin{cases} C_1(E - E_o - eU)^{1/2} & \text{for } E_o + eU < E < E_o + E_{F1} + eU = E_{F2} + eU \\ 0 & \text{otherwise} \end{cases}$$

and the number of empty states in the 'right' metal is

$$\begin{cases} C_r\, E^{1/2} & \text{for } E > E_{F2} \\ 0 & \text{otherwise} \end{cases}$$

Hence the tunnelling current $\propto \displaystyle\int_{E_{F2}}^{E_{F2}+eU} C_1(E - E_o - eU)^{1/2}\, C_r\, E^{1/2}\, dE$

$$\propto \int_{E_{F2}}^{E_{F2}+eU} (E - eU - E_o)^{1/2}\, E^{1/2}\, dE \qquad \text{(Q.E.D.)} \qquad (1)$$

(ii) When U is small, the range of integration in eqn (1) is small and it reduces to

$$(E - eU - E_o)^{1/2}\, E^{1/2}\Big|_{E=E_{E2}} (E_{F2} + eU - E_{F2}) \simeq (E_{F2} - E_0)^{1/2}\, E_{F2}^{1/2}\, eU$$

Thus the tunnelling current is proportional to U, the applied voltage. In other words, Ohm's law is satisfied.

6.11 The rule is to find five energy levels for the fermions, the sum of which is equal to 12 units and not more than two can have the same energy. Here are 10 possible distributions :

(0, 0, 1, 1, 10), (0, 0, 1, 2, 9), (0, 0, 1, 3, 8), (0, 0, 1, 4, 7), (0, 1, 1, 2, 8), (0, 1, 1, 3, 7), (0, 1, 1, 4, 6), (0, 1, 1, 5, 5), (1, 1, 2, 3, 5), (1, 1, 2, 4, 4).

7. The band theory of solids

7.2 From the free electron theory [eqn 6.16], $E_F = \frac{h^2}{2m^*}\left(\frac{3N}{8\pi}\right)^{2/3}$ (note that we replace the actual mass m by the effective mass m^*).

Now N = no. of electrons available for conduction per unit volume

= density of atoms x no. of free electrons contributed by each atom

$= \left(\frac{530}{6.94} \times 6.02\times10^{26}\right) \times 1 = 4.60\times10^{28}\ m^{-3}$

and $E_F = 4.2$ eV

Hence $m^* = \frac{h^2}{2E_F}\left(\frac{3N}{8\pi}\right)^{2/3} = 1.01\times10^{-30}$ kg or $\underline{1.11\ m_o}$

7.3 From [eqn 7.30] : $E = E_1 - 2A\cos ka$, it can be deduced that the width of the band is 4A.

The effective mass $= \hbar^2 / \frac{\partial^2 E}{\partial k^2} = \frac{\hbar^2}{2Aa^2 \cos ka}$

Hence, at the bottom of the band (k = 0), the effective mass is $\hbar^2/(2Aa^2)$ which is inversely proportional to the width of the band. (Q.E.D.)

7.4 The group velocity $V_g = \frac{1}{\hbar}\frac{\partial E}{\partial k} = \frac{1}{\hbar}\frac{\partial}{\partial k}\left(\frac{\hbar^2\alpha^2}{2m}\right)$ (from [eqn 7.5])

$= \frac{\hbar y}{a^2 m}\frac{\partial y}{\partial k}$ where $y = \alpha a$

Differentiate [eqn 7.3] with respect to k

$$\therefore \quad -a \sin ka = P \frac{y \cos y - \sin y}{y^2} \frac{dy}{dk} - \sin y \frac{dy}{dk}$$

Hence at $k = n\pi/a$, $\dfrac{dy}{dk} = \left. \dfrac{-a \sin ka}{P \dfrac{y \cos y - \sin y}{y^2} - \sin y} \right|_{k=n\pi/a} = 0$

$\therefore$ the group velocity of the electron is zero.

7.5 In three dimension, $\left(\dfrac{1}{m^*}\right)_{ij} = \dfrac{1}{\hbar^2} \dfrac{\partial^2 E}{\partial k_i \partial k_j}$

Hence the equation of motion $\dfrac{d\mathbf{v}}{dt} = \dfrac{1}{m}\mathbf{F}$ should be modified to

$$\begin{bmatrix} \dfrac{dv_x}{dt} \\ \dfrac{dv_y}{dt} \\ \dfrac{dv_x}{dt} \end{bmatrix} = \frac{1}{\hbar^2} \begin{bmatrix} \dfrac{\partial^2 E}{\partial k_x^2} & \dfrac{\partial^2 E}{\partial k_x \partial k_y} & \dfrac{\partial^2 E}{\partial k_x \partial k_z} \\ \dfrac{\partial^2 E}{\partial k_y \partial k_x} & \dfrac{\partial^2 E}{\partial k_y^2} & \dfrac{\partial^2 E}{\partial k_y \partial k_z} \\ \dfrac{\partial^2 E}{\partial k_z \partial k_x} & \dfrac{\partial^2 E}{\partial k_z \partial k_y} & \dfrac{\partial^2 E}{\partial k_z^2} \end{bmatrix} \begin{bmatrix} F_x \\ F_y \\ F_z \end{bmatrix}$$

7.6 (i) the effective mass tensor is equal to the inverse of the reciprocal mass tensor

$$= \frac{1}{a_{yy}a_{zz} - a_{yz}^2} \begin{bmatrix} \dfrac{a_{yy}a_{zz} - a_{yz}^2}{a_{xx}} & 0 & 0 \\ 0 & a_{zz} & -a_{yz} \\ 0 & -a_{yz} & a_{yy} \end{bmatrix}$$

(ii), (iii) Since $(1/m^*)_{ij} = \dfrac{\partial^2 E}{\partial k_i \partial k_j} / \hbar^2$,

$$\frac{\partial^2 E}{\partial k_x^2} = a_{xx}, \quad \frac{\partial^2 E}{\partial k_y^2} = a_{yy}, \quad \frac{\partial^2 E}{\partial k_z^2} = a_{zz}, \quad \frac{\partial^2 E}{\partial k_x \partial k_y} = \frac{\partial^2 E}{\partial k_x \partial k_z} = 0, \quad \frac{\partial^2 E}{\partial k_y \partial k_z} = a_{yz}$$

$\therefore\ E = (a_{xx}k^2 + a_{yy}k_y^2 + a_{zz}k_z^2 + a_{yz}k_yk_z)/\hbar^2$ which is an ellipsoid for constant E. (Q.E.D.)

7.7 The potential energies of the electrons in the forward- and backward-travelling wave functions are given by [eqn 7.19] :

$$V_{\pm} = \pm \frac{1}{a}\int_0^a \cos 2kx\, V(x)\, dx$$ where V(x) is the actual potential of the lattice.

For V(x) as shown in [Fig 7.2] and with 2w = a,

$$V_{\pm} = \pm \frac{2}{a}\left\{ \int_0^{a/4} -\frac{V_o}{2}\cos 2kx\, dx + \int_{a/4}^{a/2} \frac{V_o}{2}\cos 2kx\, dx \right\}$$

$$= \pm \frac{V_o}{2ka}\left(\sin\frac{3ka}{2} - \sin\frac{ka}{2} - \frac{1}{2}\sin 2ka \right)$$

In the first energy band, $k = \pi/a$ (from [eqn 7.15]). Therefore $V_{\pm} = V_o/\pi$.

The total energies of the electrons = kinetic energy + potential energy

$$= \hbar^2k^2/2m \pm V_o/\pi$$

$\therefore$ the width of the first forbidden band is $\underline{2V_o/\pi}$.

7.8 Similar to Ex 7.7,

$$V_{\pm} = \pm \frac{2}{a} \left\{ \int_{0}^{(a-w)/2} -\frac{V_o}{2} \cos 2kx \, dx + \int_{(a-w)/2}^{a/2} \frac{V_o}{2} \cos 2kx \, dx \right\}$$

$$= \pm \frac{V_o}{2ka} \left(\sin k(a + w) - \sin k(a - w) - \frac{1}{2} \sin 2ka \right)$$

For the nth energy band, $k = n\pi/a$ and as $w \longrightarrow 0$, $V_o w$ = constant,

$$V_{\pm} = \pm \lim_{w \to o} \frac{V_o}{2n\pi} \left[\sin n\pi(1 + w/a) - \sin n\pi(1 - w/a) \right]$$

$$= \pm \frac{V_o}{2n\pi} \frac{2n\pi w}{a} (-1)^n = \pm (-1)^n V_o w/a$$

$\therefore$ the total energy $= \hbar^2(n\pi/a)^2/2m \pm V_o w/a$

Thus the width of the nth allowed band is

$$\left[\hbar^2(n\pi/a)^2/2m - V_o w/a \right] - \left[\hbar^2[(n - 1)\pi/a]^2/2m + V_o w/a \right]$$

$$= h^2(2n - 1)/8ma^2 - 2V_o w/a$$

Hence the width of the allowed band increases with n, but the width of the forbidden band remains constant.

7.9 The collision time is assumed to be the same for the hole and electron. So the one with a smaller effective mass will have a higher mobility (recall that mobility $= e\tau/m^*$).

Note that $m^* = \hbar^2 / \frac{\partial^2 E}{\partial k^2}$. Since the conduction band is higher than the valence band, the width of the former is larger. Assuming that the shape of the E-k curves are similar in both bands (e.g. $E = E_1 - 2A \cos ka$ as in the Feynman's model), then $\frac{\partial^2 E}{\partial k^2}$ would be larger in the conduction band. Hence in general, an electron will have a smaller effective mass and so a higher mobility.

8. Semiconductors

8.1 The Fermi level is given by [eqn 8.24] : $E_F = \frac{E_g}{2} + \frac{3}{4}\,kT \log\left(m_h^*/m_e^*\right)$

(for derivation, see section 8.2 of the text)

∴ the difference of the Fermi level from the middle of the gap

$$= E_F - E_g/2 = \frac{3}{4}\,kT \log\left(m_h^*/m_e^*\right)$$

$$= \frac{3}{4} \times 0.025 \times \log\left(\frac{0.65}{0.067}\right) = \underline{0.043 \text{ eV}}$$

8.2 The density of electrons at an energy E is given by

$$N(E) = Z(E)F(E) \simeq C_e(E - E_g)^{1/2} \exp\left[-(E - E_F)/kT\right]$$

($\because (E - E_F)/kT >> 1$) which is maximum when $\frac{dN}{dE} = 0$

i.e. $\left(\frac{1}{2}(E - E_g)^{-1/2} - (E - E_g)^{1/2}/kT\right) \exp\left[-(E - E_F)/kT\right] = 0$

giving $E - E_g = kT/2$

∴ the most probable electron energy in the conduction band is kT/2 above the bottom of the band.

The average electron energy, $\langle E \rangle$, $= \int E\,N(E)\,dE \Big/ \int N(E)\,dE$ where the integration is from the bottom to the top of the conduction band

$$\simeq \int_{E_g}^{\infty} (E - E_g)(E - E_g)^{1/2} \exp\left(-\frac{E-E_F}{kT}\right) dE \Big/ \int_{E_g}^{\infty} (E - E_g)^{1/2} \exp\left(-\frac{E-E_F}{kT}\right) dE$$

Put $x = (E - E_g)/kT$, it becomes

$$\langle E \rangle = kT \int_0^{\infty} x^{3/2}\, e^{-x}\, dx \Big/ \int_0^{\alpha} x^{1/2}\, e^{-x}\, dx$$

But $\int_0^\infty x^{3/2} e^{-x} dx = -x^{3/2} e^{-x}\Big|_0^\infty + \int_0^\infty \frac{3}{2} x^{1/2} e^{-x} dx = \frac{3}{2}\int_0^\infty x^{1/2} e^{-x} dx$

$\therefore \langle E \rangle = 3kT/2$ (above the bottom of the conduction band)

8.3 (i) The energy at the bottom of conduction band (relative to the bottom of valence band) $= E\Big|_{k=k_1} = \frac{\hbar^2 k^2}{3m_o} = \frac{\hbar^2}{3m_o}\left(\frac{\pi}{a}\right)^2 = \frac{h^2}{12m_o a^2} = 2.54$ eV

Hence the energy gap = 2(2.54 - 2.17) = <u>0.74 eV</u> ($\because$ the Fermi level is approximately at the middle of the energy gap).

(ii) $m^* = \hbar^2 / \frac{\partial^2 E}{\partial k^2}$ [eqn 7.42]

At the bottom of conduction band $\frac{\partial^2 E}{\partial k^2} = \frac{2\hbar^2}{m_o}$

$\therefore$ the effective mass $m^* = \hbar^2 / \frac{2\hbar^2}{m_o} = \underline{m_o/2}$

8.4 The conductivity is $\sigma = N_e e\mu_e + N_h e\mu_h$ (where $N_e = N_h$) for an intrinsic semiconductor.

Now $\mu_e, \mu_h \propto T^{-3/2}$ and $N_e \propto T^{3/2}\exp(-E_g/2kT)$ (from [eqn 8.46]; $N_e = N_h$)

$\therefore 1/\rho = \sigma = A \exp(-E_g/2kT)$ where A is a constant

or $\log \rho = \frac{E_g}{2k}\frac{1}{T} - \log A$

Plot log ρ versus 1/T as shown in Fig 8.1

(i) slope $= E_g/2k = 3850$

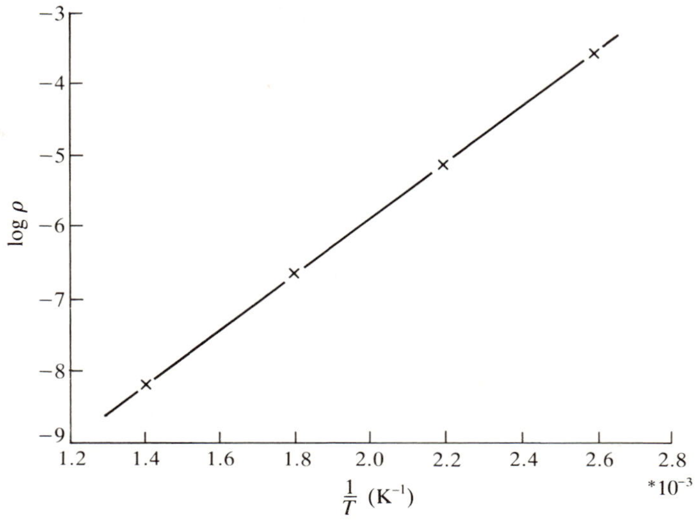

Fig 8.1 Plot of log ρ versus $1/T$.

giving E_g = <u>0.66 eV</u>

(ii) There will be optical absorption when the photon energy is larger than the energy gap of the semiconductor

i.e. when $hc/\lambda = E_g$ or $\lambda = \dfrac{6.62\mathrm{x}10^{-34}\mathrm{x}3\mathrm{x}10^{8}}{0.66\mathrm{x}1.6\mathrm{x}10^{-19}}$ = <u>1.88 μm</u>

8.5 The absorption spectrum shows a sharper rise for a direct gap material. This is because vertical transitions are easier. (To go any deeper, one should consider the probability of phonon-assisted transitions which is well beyond the scope of this course.)

8.6 Conductivity $\sigma = N_e e^2 \tau_e / m_e^* + N_h e^2 \tau_h / m_h^*$ [eqn 8.45]

From [eqn 8.46], $N_e N_h = 4\left(\frac{2\pi kT}{h^2}\right)^3 \left(m_e^* m_h^*\right)^{3/2} \exp\left(-E_g/kT\right) = K$ say, a constant at a given temperature,

$$\therefore \sigma = e^2 \left(N_e \tau_e / m_e^* + K\tau_h/(m_h^* N_e) \right)$$

It is minimum when $\frac{\partial \sigma}{\partial N_e} = 0$

i.e. when $\tau_e/m_e^* - K\tau_h/\left(m_h^* N_e^2\right) = 0$

or $N_e = \left[K\,(\tau_h/\tau_e)\cdot(m_e^*/m_h^*) \right]^{1/2}$ and $N_h = \left[K\,(\tau_e/\tau_h)\cdot(m_h^*/m_e^*) \right]^{1/2}$

$$\therefore N_h/N_e = (\tau_e/\tau_h)\cdot(m_h^*/m_e^*)$$

If $\tau_h = \tau_e$ and $m_e^*/m_h^* = \frac{1}{2}$, $N_e/N_h = \underline{0.5}$

8.7 For charge neutrality, $N_e = N_h + N_D$ assuming that all donor atoms are ionized. But $N_e N_h = N_i^2$ for semiconductors [eqn 8.47]. Rearrange the two equations to get

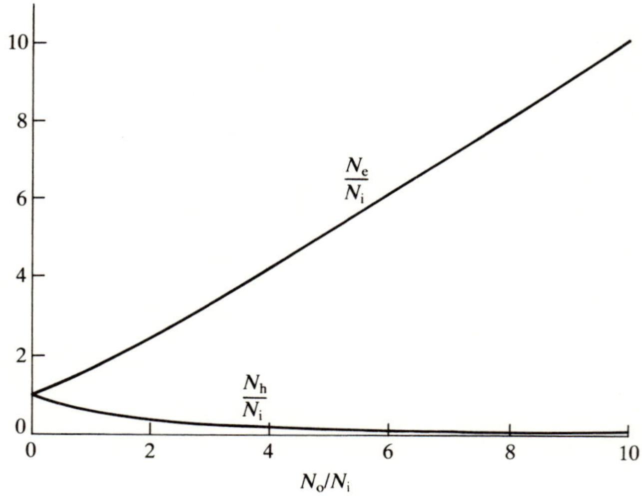

Fig 8.2 Plots of N_h/N_i and N_e/N_i as a function of N_D/N_i.

$$N_h/N_i = \frac{1}{2}\left[\left((N_D/N_i)^2 + 4\right)^{1/2} - N_D/N_i\right] \quad \text{and}$$

$$N_e/N_i = \frac{1}{2}\left[\left((N_D/N_i)^2 + 4\right)^{1/2} + N_D/N_i\right]$$

Fig 8.2 is a plot of N_h/N_i and N_e/N_i as a function of N_D/N_i in the range of $0 \leq N_D/N_i \leq 10$.

8.8 For an intrinsic semiconductor, $N_e = N_h$.

With [eqn 8.46], therefore $N_e = N_h = N_i$

where $N_i^2 = 4\left(2\pi kT/h^2\right)^3\left(m_e^*/m_h^*\right)^{3/2}\exp(-E_g/kT)$. (1)

$\therefore$ conductivity $= N_e e\mu_e + N_h e\mu_h = N_i e(\mu_e + \mu_h)$

So resistivity $\rho = \left[N_i e(\mu_e + \mu_h)\right]^{-1}$ (2)

(i) at room temperature, T = 293 K

Put in the numerical valus into eqn (1), we get $N_i = 1.54\text{x}10^{15}\ \text{m}^{-3}$

$\therefore$ from eqn (2), ρ = <u>20.3 kΩm</u>

(ii) at T = 350 K, $N_i = 6.69\text{x}10^{16}\ \text{m}^{-3}$

$\therefore$ ρ = <u>0.45 kΩm</u>

(iii) temp. coeff. $\alpha = \frac{1}{R}\frac{dR}{dT} = \frac{1}{\rho}\frac{d\rho}{dT} = -\frac{1}{N_i}\frac{dN_i}{dT}$ (from eqn (2))

$= -(3/2T + E_g/2kT^2)$ (from eqn (1)) (3)

$\therefore$ at T = 293 K (room temp.), $\alpha = \underline{-7.94\times10^{-2}\ K^{-1}}$

If we can measure the local rate of change of resistance, we can determine the temperature coefficient at that temperature. Hence, we can deduce the temperature from eqn (3).

8.9 For an n-type semiconductor, $N_e \gg N_h$ and $N_e \simeq N_D - N_A$ if all the impurity acceptors are ionized.

$\therefore$ conductivity $\simeq N_e e\mu_e \simeq (N_D - N_A)e\mu_e$

Without the impurity (i.e. $N_A = 0$), resistivity = 0.05 Ωm,

$$\therefore \frac{1}{0.05} = N_D e\mu_e \qquad (1)$$

With the impurity, resistivity = 0.06 Ωm.

$$\therefore \frac{1}{0.06} = (N_D - N_A)e\mu_e \qquad (2)$$

From eqns (1) and (2),

$$N_A = \underline{2.45\times10^{19}\ m^{-3}} \quad \text{and} \quad N_D = \underline{1.47\times10^{20}\ m^{-3}}$$

8.10 For a p-type semiconductor, $N_h \gg N_e$ and $N_h \simeq N_A$

$\therefore$ conductivity $\simeq N_h e\mu_h \simeq N_A e\mu_h$

$\therefore$ $1/10 \simeq 1.6\times10^{-19}\times0.05\ N_A$

giving $N_A = \underline{1.25\times10^{19}\ m^{-3}}$

8.11 Now $N_h \simeq N_A^- = 1.25\text{x}10^{19}\ m^{-3}$.

From [eqn 8.20], $E_F = kT \log N_v/N_h$ where $N_v = 2\left(2\pi m_h^* kT/h^2\right)^{3/2}$

Take T = 293 K, $m_h^* = 0.39\ m_o$ (from Ex 8.8), this gives $E_F = 0.33$ eV.

Now $N_A^- = N_A F(E_A)$ [eqn 8.29],

∴ fraction of acceptors which are not ionized is

$$1 - N_A^-/N_A = 1 - \left[1 + \exp(E_A - E_F)/kT\right]^{-1} = \underline{2.0\text{x}10^{-5}}$$

8.12 The density of ionized acceptors, $N_A^- = N_A\ F(E_A)$

For the Fermi level to coincide with the impurity level E_A, $F(E_A) = 1/2$,

and so $N_A^- = N_A/2$, i.e. half of the acceptor levels will be filled with electrons.

But $N_h = N_A^- + N_e$ (for charge neutrality)

∴ $N_A = 2(N_h - N_e)$

$$\text{Now} \quad N_e = 2\left(\frac{2\pi m_e^* kT}{h^2}\right)^{3/2} \exp\left(-\frac{E_g - E_F}{kT}\right) = 5.45\text{x}10^8\ \text{m}^{-3}, \text{ and}$$

$$N_h = 2\left(\frac{2\pi m_h^* kT}{h^2}\right)^{3/2} \exp\left(-\frac{E_F}{kT}\right) = 1.26\text{x}10^{22}\ \text{m}^{-3}$$

∴ impurity density $= N_A \simeq 2\text{x}1.26\text{x}10^{22} \simeq \underline{2.52\text{x}10^{22}\ \text{m}^{-3}}$

Since $N_h \gg N_e$, holes are the majority carriers in the crystal.

8.13 (i) The ionized acceptor density $N_A^- = N_A F(E_A)$

$$= 10^{23}/\left[1 + \exp(20/5 - 5)\right] \simeq \underline{7.31\text{x}10^{22}\ \text{m}^{-3}}$$

(ii) From [eqns 8.17 and 8.20],

$$N_e/N_h = \left(m_e^*/m_h^*\right)^{3/2} \exp\ (2E_F - E_g)/kT$$

$$= (0.12)^{3/2} \exp(10 - 20) = \underline{1.89\text{x}10^{-6}\ \text{m}^{-3}}$$

(iii), (iv) For charge neutrality, $N_h = N_e + N_A^-$.

With the results in parts (i) and (ii),

$$N_h = 7.31\text{x}10^{22}/(1 - 1.89\text{x}10^{-6}) \simeq \underline{7.31\text{x}10^{22}\ \text{m}^{-3}} \quad \text{and}$$

$$N_e = 7.31\text{x}10^{22}/\left[\ 1/(1.89\text{x}10^{-6}) - 1\ \right] = \underline{1.38\text{x}10^{17}\ \text{m}^{-3}}$$

(v) From [eqns 8.20 and 8.21], $N_h = 2\ \left(2\pi m_h^* kT/h^2\right)^{3/2} \exp(-E_F/kT)$

Put in the numerical values, we get $\underline{T = 171.3\ \text{K}}$

(vi) the gap energy $E_g = 20kT = 20\text{x}8.62\text{x}10^{-5}\text{x}171.3 = \underline{0.295\ \text{eV}}$

8.14 The collision time τ_e is in the form

$1/\tau_e = AT^{3/2} + B\ T^{-3/2}$ (the first term represents thermal scattering while the second term represents lattice scattering)

At $T = T_1$, the two terms are the same. Therefore $B = AT_1^3$.

Thus at $T = 2T_1$, $1/\tau_e = A(2T_1)^{3/2} + AT_1^3\ (2T_1)^{-3/2} = AT_1^{3/2}(\sqrt{8} + 1/\sqrt{8})$

The conductivity $\sigma = N_e e^2 \tau_e / m$

Now the electron density N_e increases quadratically with temperature.

$$\therefore \quad \frac{\sigma(2T_1)}{\sigma(T_1)} = \frac{(2T_1)^2 / \; AT_1^{3/2}(\surd 8 + 1/\surd 8)}{T_1^2 / \; 2AT_1^{3/2}} \quad = \underline{2.514}$$

8.15 The rate of generation of holes $= aN_eN_h$

and the rate of recombination of holes $= aN_e(N_h + \delta N_h)$

$\therefore$ the rate of change of excess holes is

$$\frac{d(\delta N_h)}{dt} = aN_eN_h - aN_e(N_h + \delta N_h) = - aN_e\delta N_h$$

Hence $\delta N_h = (\delta N_h)_0 \exp(-t/\tau_p)$ where $\tau_p = 1/(aN_e)$ (Q.E.D.)

8.16 The usual form of continuity equation is $\frac{\partial N_h}{\partial t} = -\frac{1}{e} \nabla \cdot \mathbf{J_h}$. But since there is recombination of excess holes, we need to add an extra term to describe the rate of decrease of N_h. This term, using the same arguments as in the previous example, is $(N_{hn} - N_h)/\tau_p$ where N_{hn} is the number of holes at equilibrium. Hence the continuity equation is modified

to $$\frac{\partial N_h}{\partial t} = \frac{N_{hn} - N_h}{\tau_p} - \frac{1}{e} \nabla \cdot \mathbf{J_h}$$

8.17 (i) There are four peaks in the absorption curve so one might argue that there are four types of charge carriers present.

(ii) Depends on definition. Really, there is only one type of carrier there, namely electron. All the others are artifices. Accepting however the existence of holes as separate particles, we would say that there are two types, electrons and holes. A third possibility is to maintain that since the holes have isotropic effective masses, the two kinds of holes would be regarded as two separate kinds of particles, hence there are three kinds.

(iii) Resonances occur at $B = 0.14, 0.18, 0.29, 0.44\ \text{Wbm}^{-2}$.

From [eqn 8.61], $m^* = eB/\omega_c$.

Now $\omega_c = 2\pi f_c = 2\pi \times 24000\times10^6$.

Therefore the four values of B give $m^*/m_o = 0.16, 0.21, 0.34, 0.51$ respectively.

(iv) No.

(v) From [eqn 1.66], $\dfrac{(\text{Im } k)_{\text{at resonance}}}{(\text{Im } k)_B} = 1 + (\omega - \omega_c)^2\tau^2$

Assume that Im k is plotted and zero ordinate means Im k = 0. Then read from [Fig 8.14] Im k at, say, half of its resonance value. The equation to be solved is then

$$2 = 1 + (\omega - \omega_c)^2\tau^2$$

giving $\tau = \left[\ \omega\ |1 - B/B_r|\ \right]^{-1} \simeq 4.6\times10^{-11}$ s and 5.8×10^{-11} s ($B = 0.12, 0.49\ \text{Wbm}^{-2}$ for the two holes).

(vi) Power absorbed at resonance is proportional to $\omega_p^2\tau$ (from [eqn

1.66]). Now since the two hole peaks are about the same, it follows that

$$\omega_{p1}^2 \tau_1 = \omega_{p2}^2 \tau_2$$

With [eqn 1.53], therefore $N_{h1}\tau_1/m_1^* = N_{h2}\tau_2/m_2^*$,

whence $N_{h1}/N_{h2} = (\tau_2/\tau_1)\cdot(m_1^*/m_2^*) = (\tau_2/\tau_1)\cdot(B_{r1}/B_{r2})$

$$= (6.5/5)\cdot(0.14/0.44) = \underline{0.41}$$

(vii) It is because the deep level is sparsely populated.

9. Principles of semiconductor devices

9.1 From [eqn 8.17],

$$\begin{cases} N_{en} = N_c \exp\left[-(E_g - E_{Fn})/kT\right] \\ N_{ep} = N_c \exp\left[-(E_g - E_{Fp})/kT\right] \end{cases}$$

where E_{Fn}, E_{Fp} are the Fermi levels in the n- and p-type materials respectively; N_c is a constant given by [eqn 8.18].

Divide the two equations to get $N_{en}/N_{ep} = \exp\left[(E_{Fn}-E_{Fp})/kT\right]$

However, the 'built-in' voltage = difference in Fermi level / e

$$\text{Hence, } U_o = (E_{Fn}-E_{Fp})/e = \frac{kT}{e} \log \frac{N_{en}}{N_{ep}} \qquad \text{(Q.E.D.)}$$

Similarly, for holes, from [eqn 8.20]

$$\begin{cases} N_{hn} = N_v \exp\left(- E_{Fn}/kT\right) \\ N_{hp} = N_v \exp\left(- E_{Fp}/kT\right) \end{cases}$$

$$\therefore\ U_o = (E_{Fn}-E_{Fp})/e = \frac{kT}{e} \log \frac{N_{hp}}{N_{hn}} \qquad \text{(Q.E.D.)}$$

9.2 At thermal equilibrium, $J_e = 0$

$$\therefore\ e\mu_e N_e \mathscr{E} + eD_e \frac{dN_e}{dx} = 0$$

$$-\int \mathscr{E}\, dx = \frac{D_e}{\mu_e} \int \frac{dN_e}{N_e} \qquad (1)$$

Integrate across the transition region.

R.S. of eqn (1) gives $\frac{D_e}{\mu_e}\log(N_{en}/N_{ep})$ where N_{en}, N_{ep} are the electron densities beyond the transition region in the n-type and p-type materials respectively.

Note that $-\int \mathscr{E}\,dx$ across the transition region is the 'built-in' voltage.

Thus $U_o = \frac{D_e}{\mu_e}\log\frac{N_{en}}{N_{ep}}$ (Q.E.D.)

Similarly, by taking $J_h = 0$, we obtain

$$U_o = \frac{D_h}{\mu_h}\log\frac{N_{hp}}{N_{hn}}$$

By comparing the results with that in Ex 9.1, we get

$\frac{D_e}{\mu_e} = \frac{D_h}{\mu_h} = \frac{kT}{e}$ which is the Einstein relationship. (Q.E.D.)

Eliminate $\mathscr{E}$ from the expressions of J_e and J_h, and note that J_e and J_h are separately zero,

$$\therefore \frac{D_e}{\mu_e}\frac{dN_e}{N_e} = -\frac{D_h}{\mu_h}\frac{dN_h}{N_h}$$

But $D_e/\mu_e = D_h/\mu_h$

$$\therefore \frac{dN_e}{N_e} + \frac{dN_h}{N_h} = 0$$

$\log N_e + \log N_h =$ constant or $N_eN_h =$ constant (Q.E.D.)

9.3 The density gradient of impurities implies a spatially varying electron density and a resulting diffusion current. Therefore $\sigma = N_e e\mu_e$.

At thermal equilibrium, conduction current = diffusion current,

$\therefore\ N_e e\mu_e \mathscr{E} = -eD_e \dfrac{dN_e}{dx}$ where $\mathscr{E}$ is the local electric field

$$-\int \mathscr{E}\,dx = \frac{D_e}{\mu_e}\int \frac{dN_e}{N_e}$$

Integrate across the semiconductor,

$\therefore$ the 'built-in' voltage $U_o = \dfrac{D_e}{\mu_e}\log\dfrac{N_{e2}}{N_{e1}}$ where N_{e1}, N_{e2} are the electron densities at the low and high impurity ends respectively.

Rearrange to get $N_{e1}e\mu_e = N_{e2}e\mu_e \exp(-U_o\mu_e/D_e)$

$\therefore\ \sigma_1 = \sigma_2 \exp(-U_o e/kT)$ (from the Einstein relationship)

or $\rho_1 = 10 \exp\left[0.125\text{x}1.6\text{x}10^{-19}/(1.38\text{x}10^{-23}\text{x}293)\right]\ \Omega\text{m} = \underline{1.4\ \text{k}\Omega\text{m}}$

9.4 In the semiconductor (see Fig 9.1)

$$\frac{d^2U}{dx^2} = -\frac{eN_D}{\varepsilon_s\varepsilon_o} \quad \text{(Poisson's equation)}$$

Integrate once, $\mathscr{E} = -\dfrac{dU}{dx} = \dfrac{eN_D}{\varepsilon_s\varepsilon_o}x + \mathscr{E}_o$ where $\mathscr{E}_o$ is the electric field at the insulator-semiconductor boundary.

Suppose the width of the depletion region is x_n. Then at $x = x_n$, $\mathscr{E}$ is zero because there is no charge imbalance to the right of x_n.

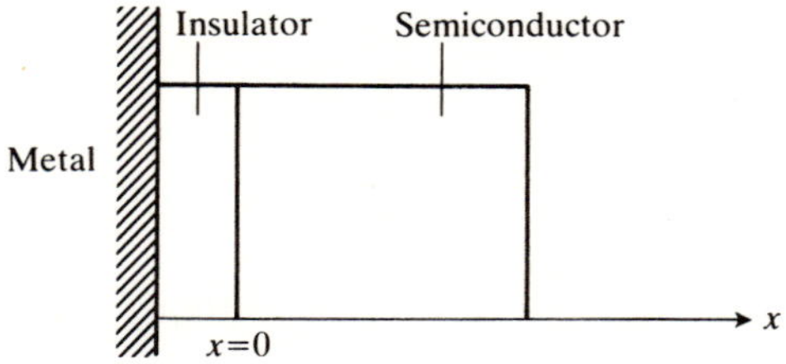

Fig 9.1 The metal-insulator-semiconductor junction.

Hence $\mathscr{E}_o = -\dfrac{eN_D}{\varepsilon_s\varepsilon_o} x_n$.

The potential difference across the semiconductor is

$$-\int_0^{x_n} \mathscr{E}\,dx = -\frac{eN_D}{\varepsilon_s\varepsilon_o}\left(x_n^2/2 - x_n^2\right) = \frac{eN_D x_n^2}{2\varepsilon_s\varepsilon_o}$$

The electric field in the insulator is $(\varepsilon_s/\varepsilon_i)\mathscr{E}_o$

$\therefore$ the potential drop across it $= (\varepsilon_s/\varepsilon_i)\mathscr{E}_o d_i = \dfrac{eN_D x_n}{\varepsilon_i\varepsilon_o} d_i$

So the total voltage drop $= U_o = \dfrac{eN_D x_n}{\varepsilon_i\varepsilon_o} d_i + \dfrac{eN_D x_n^2}{2\varepsilon_s\varepsilon_o}$

$$\therefore\ x_n^2 + \frac{2\varepsilon_s d_i}{\varepsilon_i} x_n - \frac{2\varepsilon_s\varepsilon_o U_o}{eN_D} = 0$$

giving $x_n = -(\varepsilon_s/\varepsilon_i)d_i + \left[(\varepsilon_s/\varepsilon_i)^2 d_i^2 + 2\varepsilon_s\varepsilon_o U_o/eN_D\right]^{1/2}$

9.5 (i) When $d < d_o$, the space charge density is given by

$$\rho = \begin{cases} \dfrac{2eN_D}{d_o} x & \text{for } 0 < x < d/2 \\ 0 & \text{for } x > d/2 \end{cases}$$

$\therefore$ the electric field $\mathscr{E} = \int \rho/\varepsilon\, dx = \left[x^2 - (d/2)^2\right] eN_D/\varepsilon d_o$

Thus the 'built-in' voltage $U_o = -2\int_0^{d/2} \mathscr{E}\,dx = eN_D d^3/6\varepsilon d_o$ ($\because$ by symmetry,

the voltage drop from $-d/2$ to 0 is the same as that from 0 to $d/2$)

Hence $d = \left[6U_o\varepsilon d_o/eN_D\right]^{1/3}$

(ii) When $d > d_0$,

$$\rho = \begin{cases} 0 & \text{for } x > d/2 \\ eN_D & \text{for } d_0/2 < x < d/2 \\ (2eN_D/d_0)\,x & \text{for } 0 < x < d_0/2 \end{cases}$$

$$\therefore\ \mathscr{E} = \int \rho/\varepsilon\, dx$$

$$= \begin{cases} eN_D(x - d/2)/\varepsilon & \text{for } d_0/2 < x < d/2 \\ \dfrac{eN_D}{\varepsilon d_0}\left(x^2 - (d_0/2)^2 \right) + eN_D(d_0 - d)/2\varepsilon & \text{for } 0 < x < d_0/2 \end{cases}$$

Note that the integration constants are determined by

(i) the electric field at $x = d/2$ is zero

(ii) matching the electric field at $x = d_0/2$

Thus $$\frac{U_0}{2} = -\int_0^{d/2} \mathscr{E}\, dx$$

$$= -\int_0^{d_0/2} \left\{ \frac{eN_D}{\varepsilon d_0}\left[x^2 - (d_0/2)^2\right] + \frac{eN_D(d_0 - d)}{2\varepsilon} \right\} dx - \int_{d_0/2}^{d/2} \frac{eN_D}{\varepsilon}(x - d/2)\, dx$$

$$= \frac{eN_D}{\varepsilon}\left(d^2/8 - d_0^2/24 \right)$$

or $$d = \left(4U_0\varepsilon/eN_D + d_0^2/3 \right)^{1/2}$$

9.6 Since $N_{en} = \sigma_n/e\mu_e$, $N_{hp} = \sigma_p/e\mu_h$ and $N_{hn}N_{en} = N_i^2$ (1)

and from the expression given in Ex 9.1,

$$U_o = \frac{kT}{e}\log\frac{N_{hp}}{N_{hn}} = \frac{kT}{e}\log\left(\frac{\sigma_p\sigma_n}{N_i^2 e^2\mu_e\mu_h}\right)$$

Substitute the values into the expression,

$$U_o = \begin{cases} 0.35\ \text{V} & \text{for Ge} \\ 0.77\ \text{V} & \text{for Si} \end{cases}$$ (taking room temperature as 293 K)

9.7 The density of holes will increase from the equilibrium density by a factor of exp eV/kT when a forward bias voltage V is applied,

i.e. $N_{h,injected} = N_{hn}\exp eV/kT$

$$= \frac{N_i^2 e\mu_e}{\sigma_n}\exp eV/kT$$ (from eqn (1) of Ex 9.6)

$$= \underline{1.81\text{x}10^{19}\ \text{m}^{-3}}$$

Similarly, $N_{e,injected} = N_{ep}\exp eV/kT = \frac{N_i^2 e\mu_h}{\sigma_p}\exp eV/kT = \underline{8.55\text{x}10^{16}\ \text{m}^{-3}}$

9.8 (i) $N_D^+ = N_D\left[1 - F(E_D)\right]$

Now $N_D^+ = \alpha N_D$

$$\therefore\ \alpha = 1 - F(E_D) = 1 - \left[1 + \exp(E_D - E_F)/kT\right]^{-1}$$

$$\therefore\ \exp(E_D - E_F)/kT = \frac{\alpha}{1-\alpha}$$

or $\underline{E_F - E_D = kT\log\left(\frac{1-\alpha}{\alpha}\right)}$

(ii) $N_A^- = N_A F(E_A)$

But $N_A^- = \beta N_A$,

$$\therefore \beta = F(E_A) = \left[1 + \exp(E_A - E_F)/kT \right]^{-1}$$

$$\text{or} \quad \exp(E_A - E_F)/kT = \frac{1 - \beta}{\beta}$$

$$\underline{E_F - E_A = kT \log\left(\frac{\beta}{1 - \beta}\right)}$$

(iii) the 'built-in' voltage = difference in the Fermi levels

$$= \left[E_D + kT \log\left(\frac{1 - \alpha}{\alpha}\right) \right] - \left[E_A + kT \log\left(\frac{\beta}{1 - \beta}\right) \right]$$

$$\therefore 1.05 = 1.1 + kT \log\left(\frac{1 - 0.5}{0.5}\right) - 0.1 - kT \log\left(\frac{0.05}{1 - 0.05}\right)$$

giving $T = \underline{197\ K}$

9.9 From [eqns 9.5 and 9.6],

$$\mathscr{E} = \begin{cases} -\dfrac{eN_A}{\varepsilon}(x - x_p) & \text{in the p-region} \\ \dfrac{eN_D}{\varepsilon}(x - x_n) & \text{in the n-region} \end{cases}$$

$$\therefore \mathscr{E}_{max} = \frac{eN_A}{\varepsilon} x_p = \frac{eN_D}{\varepsilon} x_n$$

$$\text{or} \quad x_p = \frac{\varepsilon \mathscr{E}_{max}}{eN_A} \quad \text{and} \quad x_n = \frac{\varepsilon \mathscr{E}_{max}}{eN_D}$$

From [eqn 9.8], $U_o = \frac{e}{2\varepsilon}\left(N_A x_p^2 + N_D x_n^2 \right)$

$$\therefore U_o = \frac{\varepsilon \mathscr{E}_{max}^2}{2e}\left(\frac{1}{N_A} + \frac{1}{N_D} \right) \qquad (1)$$

$$= \frac{16\text{x}8.85\text{x}10^{-12}\text{x}(2\text{x}10^{7})^{2}}{2\text{x}(1.6\text{x}10^{-19})} \times (10^{-23} + 10^{-22}) = \underline{19.47\ \text{V}}$$

9.10 The continuity equation is

$$\frac{\partial N_h}{\partial t} = -\frac{N_h - N_{hn}}{\tau_p} - \frac{1}{e}\nabla\cdot\mathbf{J_h}$$ where **J** is the current density vector.

Now the diffusion current dominates, therefore $J_h \simeq -eD_h \frac{\partial N_h}{\partial x}$

Assume that $\frac{\partial}{\partial t} = 0$,

$$\therefore\ 0 = \frac{N_{hn} - N_h}{\tau_p} + D_h \frac{\partial^2 N_h}{\partial x^2}$$

yielding $N_h - N_{hn} = A \exp\left[-x/(D_h\tau_p)^{1/2}\right] + B \exp\left[x/(D_h\tau_p)^{1/2}\right]$

where A, B are constants.

When $x \longrightarrow \infty$, $N_h \longrightarrow N_{hn}$, therefore B = 0

At x = 0 (chosen to be in the n-type material, at the end of the transition region), N_h = injected hole density = $N_{hn}\exp(eU_1/kT)$.

Hence $N_h - N_{hn} = N_{hn}\left[\exp eU_1/kT - 1\right]\exp\left(-x/(D_h\tau_p)^{1/2}\right)$

9.11 From the result of Ex 9.10, the injected hole density is reduced by a factor of e when $x = \sqrt{D_h\tau_p} = \sqrt{(0.0044\text{x}200\text{x}10^{-6})}$ m = $\underline{0.94\ \text{mm}}$

9.12 Neglect the conduction current.

$$J_h = -eD_h \frac{\partial N_h}{\partial x} \quad (N_h \text{ is given by the result in Ex 9.10})$$

$$= e\left(\frac{D_h}{\tau_p}\right)^{1/2} N_{hn}\left[\exp eU_1/kT - 1\right]\exp\left(-x/(D_h\tau_p)^{1/2}\right)$$

9.13 By analogy to the expression of J_h in Ex 9.12, the current injected into the p-type material is

$$J_e = e\left(\frac{D_e}{\tau_n}\right)^{1/2} N_{ep}\left[\exp eU_1/kT - 1\right]$$

Hence the total injected current,

$$J_e + J_h = e\left[(D_e/\tau_n)^{1/2} N_{ep} + (D_n/\tau_p)^{1/2} N_{hn}\right]\left[\exp eU_1/kT - 1\right]$$

Thus $I_o = e\left[(D_e/\tau_n)^{1/2} N_{ep} + (D_n/\tau_p)^{1/2} N_{hn}\right]$ x cross-sectional area

9.14 (i) conductivity $\sigma = N_e e\mu_e + N_h e\mu_h$

For an intrinsic semiconductor, $N_e = N_h$.

So with [eqn 8.46], $N_e = N_h = 2\left(2\pi kT/h^2\right)^{3/2}\left(m_e^* m_h^*\right)^{3/4} \exp(-E_g/2kT)$

Hence $\sigma = \sigma_o \exp(-E_g/2kT)$ where $\sigma_o = 2e(\mu_e + \mu_h)\left(2\pi kT/h^2\right)^{3/2}\left(m_e^* m_h^*\right)^{3/4}$

(ii) The percentage change in conductivity when the pressure increases from atmospheric pressure to 10 atmospheres is

$$\left[\frac{\exp\left[-(1.1 - 2\text{x}10^{-3}\text{x}10)/2\text{x}0.025\right]}{\exp(-1.1/2\text{x}0.025)} - 1\right] \text{x } 100\ \% = \underline{49.2\ \%}$$

(iii) From Ex 9.13,

$$I_o = e\left[(D_e/\tau_n)^{1/2}N_{ep} + (D_n/\tau_p)^{1/2}N_{hn}\right] \times \text{cross-sectional area}$$

Now $1/\tau_p$ (the recombination constant of holes in the p-type material) is proportional to N_{en} and $1/\tau_n$ is proportional to N_{hp}. Therefore

$$I_o = AN_{ep}\sqrt{N_{hp}} + BN_{hn}\sqrt{N_{en}} \quad \text{where A, B are constants.}$$

From [eqns 8.17 and 8.20],

$$N_{ep}\sqrt{N_{hp}} \propto \exp\left[(E_{Fp} - 2E_g)/2kT\right] \text{ and } N_{hn}\sqrt{N_{en}} \propto \exp\left[-(E_g + E_{Fn})/2kT\right]$$

If the temperature is not too high, $E_{Fp} << E_g$ (assuming that the acceptor level is close to the valence band) and $E_{Fn} \simeq E_g$ (assuming that the donor level is close to the conduction band) (see [Figs 8.5 and 8.6]). Therefore both $N_{ep}\sqrt{N_{hp}}$ and $N_{hn}\sqrt{N_{en}} \propto \exp(-E_g/kT)$ and hence I_o is approximately proportional to $\exp(-E_g/kT)$.

(iv) For the same pressure change, I_o increases by 122.6 %. Thus I_o is much more sensitive to pressure change than that of σ.

9.15 The energy diagram of the p-n junction looks like Fig 9.2. Measure the energy from the top of the valence band of the n-material.

Let $E_c(x)$ be the energy at the bottom of the conduction band.

$$N_e = N_c \exp\left(-\frac{E_c(x) - E_F}{kT}\right)$$

Before joining the two semiconductors,

$$N_e = \begin{cases} N_c \exp\left(-\dfrac{E_g - E_F}{kT}\right) & \text{in the n-material } (x > 0) \\ N_c \exp\left(-\dfrac{eU_o + E_g - E_F}{kT}\right) & \text{in the p-material } (x < 0) \end{cases}$$

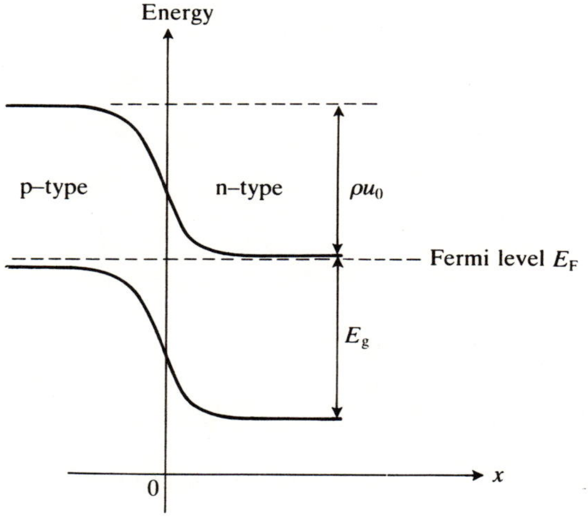

Fig 9.2 The energy diagram of the p-n junction.

Thus the increase in the number of electrons after joining the two semiconductors is

$$\begin{cases} N_c \exp\left(-\dfrac{E_g - E_F}{kT}\right)\left[\exp\left(-\dfrac{E_g - E_c(x)}{kT}\right) - 1\right] & \text{at } x > 0 \\ N_c \exp\left(-\dfrac{eU_o + E_g - E_F}{kT}\right)\left[\exp\left(-\dfrac{E_g + eU_o - E_c(x)}{kT}\right) - 1\right] & \text{at } x < 0 \end{cases}$$

Since the junction is symmetric, the $E_c(x)$ curve must be antisymmetric with respect to $E_c(0)$.

$$\therefore E_c(0) - E_c(x_o) = E_c(-x_o) - E_c(0)$$

But $E_c(0) = E_g + \frac{1}{2} eU_o$

$$\therefore E_c(-x_o) = 2E_g + eU_o - E_c(x_o)$$

Hence the net decrease in the number of electrons at $x=-x_o$ and $x=x_o$ is

$$\text{net decrease} = N_c \exp\left(-\frac{E_g - E_F}{kT}\right)\left\{1 - \exp\left(\frac{E_g - E_c(x_o)}{kT}\right)\right\}$$

$$+ N_c \exp\left(-\frac{eU_o + E_g - E_F}{kT}\right)\left\{1 - \exp\left(-\frac{E_g - E_c(x_o)}{kT}\right)\right\}$$

$$= 2N_c \exp\left(-\frac{E_g - E_F}{kT} - \frac{eU_o}{2kT}\right)\left\{\cosh\frac{eU_o}{2kT} - \cosh\left(\frac{eU_o}{2kT} - \frac{E_c - E_g}{kT}\right)\right\}$$

Note that $0 < E_c - E_g < eU_o/2$ for all $x_o > 0$,

$\therefore$ net decrease > 0

We have proved that the loss of electrons at a distance x_o in the n-type material is larger than the gain of electrons at a distance $-x_o$ in the p-type material. Thus the total loss of electrons from the n-side, $\int_0^L \Delta N_e(x)\,dx$, must be larger than the total gain of electrons on the p-side, $\int_{-L}^0 \Delta N_e(x)\,dx$.

Similar treatment will prove that the gain of holes from the n-side is less than the loss of holes on the p-side.

Hence the total number of carriers is reduced when the two samples are joined together.

9.16 (i) Rectifier equation : $I = I_o\left[\exp eU_1/kT - 1\right]$ [eqn 9.17]

When it is forward biased, $\exp eU_1/kT \gg 1$,

$\therefore I = I_o \exp eU_1/kT$

Now when $U_1 = 1.3$ V, $I = 0.015$ mA and when $U_1 = 1.6$ V, $I = 17$ mA

$$\therefore \begin{cases} 0.015\times10^{-3} = I_o \exp 1.3e/kT \\ 17\times10^{-3} = I_o \exp 1.6e/kT \end{cases}$$

Solving to get $I_o \simeq \underline{8.7\times10^{-19}\ \text{mA}}$ and $T \simeq \underline{495\ \text{K}}$

(ii) From eqn (1) of Ex 9.9, the breakdown voltage U_{br} is given by

$$U_{br} = \frac{\varepsilon\mathscr{E}^2_{max}}{2e}\left(\frac{1}{N_A} + \frac{1}{N_D}\right)$$

Since the junction is heavily doped on the n-side, $N_D >> N_A$

$$\therefore\ U_{br} \simeq \frac{\varepsilon\mathscr{E}^2_{max}}{2eN_A}$$

Put in the numerical values into it, we get $N_A = \underline{1.1\times10^{23}\ \text{m}^{-3}}$ and assuming complete ionization, $N_h = N_A$.

(iii) The capacitance of a p-n junction (per unit area) is given by

$$C = \left\{\frac{\varepsilon e}{2(U_o + U_1)}\ \frac{N_A N_D}{N_A + N_D}\right\}^{1/2} \qquad \text{[eqn 9.21]}$$

As $N_D >> N_A$, $$C^2 \simeq \frac{\varepsilon e N_A}{2(U_o + U_1)} \qquad (1)$$

When $U_1 = 1$ V, CA = 83.1 pF and when $U_1 = 10$ V, CA = 41.7 pF where A is the junction area.

Thus $$\left(\frac{83.1}{41.7}\right)^2 = \frac{U_o + 10}{U_o + 1}$$

$\therefore$ the 'built-in' voltage $U_o = \underline{2.0\ \text{V}}$

Substitute into eqn (1) to get the junction area $A = \underline{1.49\times10^{-7}\ \text{m}^2}$

10. Dielectric materials

10.2 From [eqn 10.33], $\alpha = \frac{\varepsilon' - 1}{\varepsilon' + 2} \cdot \frac{3\varepsilon_o}{N_m}$ (Clausius-Mosotti equation)

At 273 K, 1 atmosphere, argon may be regarded as an ideal gas. Hence the number of atoms per $m^3 = 6.02\text{x}10^{26}\text{x}1.013\text{x}10^5/(8.31\text{x}10^3\text{x}273) = 2.69\text{x}10^{25}$

Now $\varepsilon' - 1$ = susceptibility $= 4.35\text{x}10^{-4}$

$$\therefore \text{ polarizability } \alpha = \frac{4.35\text{x}10^{-4}}{3 + 4.35\text{x}10^{-4}} \text{ x } \frac{3\text{x}8.85\text{x}10^{-12}}{2.69\text{x}10^{25}} = \underline{1.43\text{x}10^{-40}\ \text{Fm}^2}$$

10.3 The effective electric field is given by [eqn 10.32] :

$$\mathscr{E}' = \frac{1}{3}(\varepsilon' + 2)\,\mathscr{E}$$

Rearrange the Clausius-Mosotti equation [eqn 10.33] to get

$$\varepsilon' = \frac{3\varepsilon_o/N_m + 2\alpha}{3\varepsilon_o/N_m - \alpha}$$

With $\alpha = 10^{-40}\ \text{Fm}^2$ and $N_m = 5\text{x}10^{28}\ \text{m}^{-3}$, we get $\varepsilon' = 1.70$

Thus $\mathscr{E}' = \frac{1}{3}(1.7 + 2) \text{ x } 1 = \underline{1.23\ \text{Vm}^{-1}}$

10.4 The force on the charge centres of the dipole are both $q\mathscr{E}$ but in opposite directions (see Fig 10.1). Therefore the torque on the dipole is

$$2q\mathscr{E}\frac{d}{2}\sin\theta = \mu\mathscr{E}\sin\theta \quad (\because \mu = qd)$$

Thus the energy of the dipole $= \int \mu\mathscr{E}\sin\theta\, d\theta = -\mu\mathscr{E}\cos\theta$ (taking the energy at $\theta = \pi/2$ to be zero). (Q.E.D.)

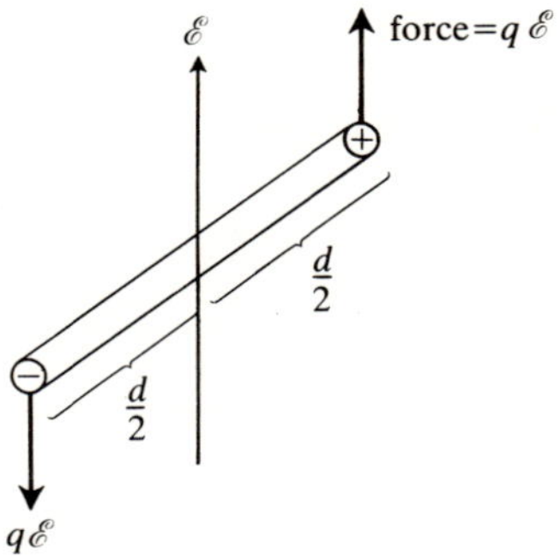

Fig 10.1 The forces on a dipole due to an electric field.

10.5 From the Debye equations [eqns 10.26 & 10.27], the loss tangent can be written as

$$\tan\delta = \varepsilon''/\varepsilon' = \frac{\omega\tau(\varepsilon_s - \varepsilon_\infty)}{\varepsilon_s + \varepsilon_\infty\omega^2\tau^2} \quad \text{(note that } \tau \text{ is temperature dependent)}$$

$\tan\delta$ is maximum when $\tau = \frac{1}{\omega}(\varepsilon_s/\varepsilon_\infty)^{1/2}$ and its value is

$$\Big[\tan\delta\Big]_{max} = \frac{\varepsilon_s - \varepsilon_\infty}{2\sqrt{\varepsilon_s\varepsilon_\infty}} = 0.0851$$

Since the maxima of both curves are not far from 0.0851, it seems that orientation polarization as expressed by the Debye equations is responsible for the temperature variation of τ.

Take the maxima of the two curves at

(i) $T_1 = 516$ K, frequency = 695 Hz

(ii) $T_2 = 574.5$ K, frequency = 6950 Hz

At the maxima, $\tau = \frac{1}{\omega}(\varepsilon_s/\varepsilon_\infty)^{1/2}$. So for $\tau = \tau_0 \exp(H/kT)$, we have

$$H = \frac{kT_1T_2}{T_2 - T_1}\log(\omega_2/\omega_1) = \underline{1.01 \text{ eV}}$$

and $\tau_0 = \frac{1}{\omega_1}(\varepsilon_s/\varepsilon_\infty)^{1/2}\exp(-H/kT_1) = \underline{3.77\text{x}10^{-14}\text{ s}}$

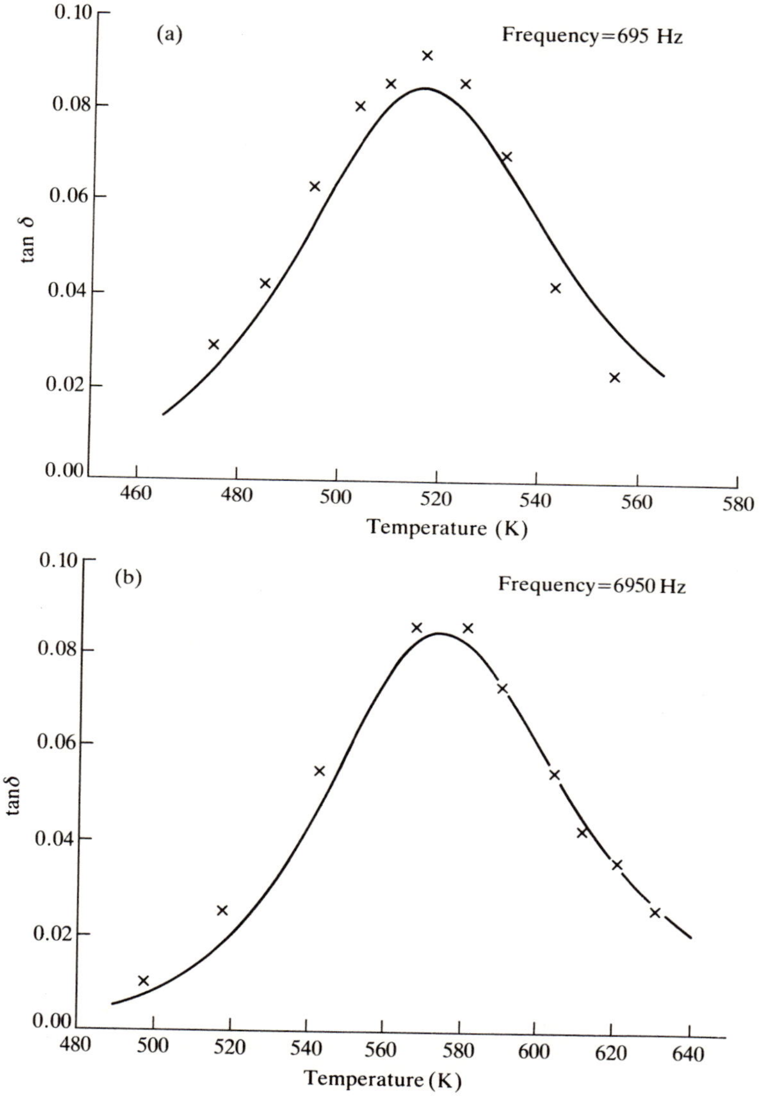

Fig 10.2 Plot of tan δ as a function of temperature T at (a) 695 Hz, (b) 6950 Hz

With these values of H and τ, the corresponding curves for f = 695 Hz, 6950 Hz are shown in Fig 10.2 a, b; the measured points are indicated by the symbols x.

At 500 K, $\tau \simeq 5.8\times10^{-4}$ s. The electric flux density will decay exponentially with this as the time constant when the steady electric field is suddenly removed.

10.6 $D + a\dfrac{dD}{dt} = b\mathscr{E} + c\dfrac{d\mathscr{E}}{dt}$

Take the Fourier transform (to get the frequency response) on both sides,

$\therefore\ (1 - i\omega a)\, D(\omega) = (b - i\omega c)\, \mathscr{E}(\omega)$

But $D(\omega) = \varepsilon(\omega)\mathscr{E}(\omega)$ where $\varepsilon(\omega)$ is the permittivity given by the Debye equation [eqn 10.25].

$$\therefore\ \frac{b - i\omega c}{1 - i\omega a} = \varepsilon_\infty + \frac{\varepsilon_s - \varepsilon_\infty}{-i\omega\tau + 1} = \frac{\varepsilon_s - i\omega\tau\varepsilon_\infty}{-i\omega\tau + 1}$$

Hence $\underline{a = \tau,\ b = \varepsilon_s,\ c = \tau\varepsilon_\infty}$.

Alternatively, one may argue as follows : (i) at zero frequency, $d/dt = 0$, hence $b = \varepsilon_s$, (ii) the flux density must decay with a time constant τ, hence $a = \tau$, (iii) at very high frequencies $-ia\omega D = -ic\omega\mathscr{E}$, hence $\varepsilon_\infty = c/a = c/\tau$.

10.8 (a) The capacitor would be equivalent to two capacitors joining in parallel as shown in Fig 10.3a.

For capacitor 1,

$C_1 = \varepsilon_0\varepsilon_r\cdot\frac{2}{3}\, A/d$ where d is the thickness of the the dielectric

For capacitor 2, it can be regarded as two capacitors in series as shown in Fig 10.3b.

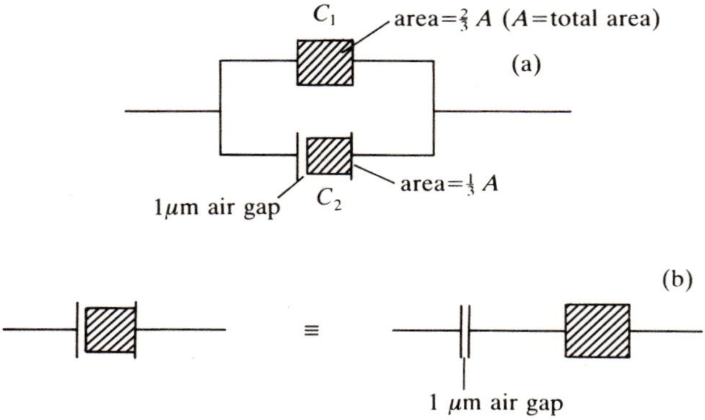

Fig 10.3 (a) An equivalent circuit of the capacitor.

(b) An equivalent circuit of the capacitor with an air gap.

Thus $C_2 = \frac{\varepsilon_o\varepsilon_r A}{3d} \Big/ \left(\frac{\varepsilon_r d_a}{d} + 1\right)$ where d_a is the length of the air gap = 1 μm

Hence the overall capacitance

$$= \frac{\varepsilon_o\varepsilon_r A}{d}\left[\frac{2}{3} + \frac{1}{3}\left(\frac{\varepsilon_r d_a}{d} + 1\right)^{-1}\right] = 0.78\ \frac{\varepsilon_o\varepsilon_r A}{d}$$

Without the air gap, the capacitance is $\frac{\varepsilon_o\varepsilon_r A}{d}$. Therefore the capacitance is reduced by about 22 % when there is the air gap.

(b) Assume that the breakdown begins at air. Then at breakdown, the electric field at the air gap is 3 MVm^{-1}. Therefore the electric field in the dielectric is $3/\varepsilon_r$ MVm^{-1} = 3 kVm^{-1} which is smaller than the breakdown field of the dielectric. So our assumption is correct.

the breakdown voltage = voltage across the air gap + voltage drop across the dielectric

$$= 3\times10^6\times1\times10^{-6} + \frac{3\times10^6}{1000}\times0.5\times10^{-3} = 4.5\ V$$

Without the air gap, the breakdown voltage is $2\times10^6\times0.5\times10^{-3} = 1000$ V. Hence the breakdown voltage drops enormously when there is an air gap.

10.9 The relevant equations are

$$\left\{\begin{aligned}
T &= cS - e\mathcal{E} \quad &&[\text{eqn } 10.34]\\
D &= \varepsilon\mathcal{E} + eS \quad &&[\text{eqn } 10.35]\\
J &= N_e e\mu\mathcal{E} + eD_e\frac{\partial N_e}{\partial z} \quad &&(\text{from Ex } 9.2)\\
\frac{\partial N_e}{\partial t} &= \frac{1}{e}\frac{\partial J}{\partial z} \quad &&(\text{continuity equation})\\
\frac{\partial D}{\partial z} &= -e(N_e + N_+) \quad &&(\text{Maxwell equation})\\
\frac{\partial^2 T}{\partial z^2} &= \rho_m\frac{\partial^2 S}{\partial t^2} \quad &&(\text{wave equation})
\end{aligned}\right. \qquad (1)$$

where N_+ is the density of the background ions.

Let $\xi = \xi_0 + \xi_1\exp\left[-i(\omega t - kz)\right]$ where $\xi = \mathcal{E}$, D, T, S, J, N_e and assume that $\xi_0 \gg \xi_1$ (i.e. the d.c. quantities are much larger than the a.c. quantities).

Substitute into eqn (1), neglect the products of a.c. quantities and equate the d.c. terms and a.c. terms separately.

$$\therefore\left\{\begin{aligned}
T_0 &= cS_0 - e\mathcal{E}_o\\
D_0 &= \varepsilon\mathcal{E}_0 + eS_o\\
J_0 &= N_{eo}e\mu\mathcal{E}_o\\
0 &= -e(N_{eo} + N_+)
\end{aligned}\right. \qquad (2)$$

and

$$\begin{cases} T_1 = cS_1 - e\mathscr{E}_1 \\ D_1 = \varepsilon\mathscr{E}_1 + eS_1 \\ J_1 = N_{eo}e\mu\mathscr{E}_1 + N_{e1}e\mu\mathscr{E}_o + ikeD_eN_{e1} \\ -i\omega eN_{e1} = ikJ_1 \\ ikD_1 = -eN_{e1} \\ k^2T_1 = \rho_m\omega^2S_1 \end{cases} \tag{3}$$

Eliminate T_1, J_1 and D_1 in eqn (3),

$$\begin{cases} (c - \rho_m\omega^2/k^2)\, S_1 - e\mathscr{E}_1 = 0 \\ eS_1 + \varepsilon\mathscr{E}_1 + eN_{e1}/ik = 0 \\ N_{eo}e\mu\mathscr{E}_1 + (e\mu\mathscr{E}_o + ikeD_e + \omega e/k)\, N_{e1} = 0 \end{cases}$$

Eliminate S_1 to get

$$\begin{cases} \left[\varepsilon + e^2/(c - \rho_m\omega^2/k^2)\right] \mathscr{E}_1 + eN_{e1}/ik = 0 \\ N_{eo}\mu\mathscr{E}_1 + (\mu\mathscr{E}_o + ikD_e + \omega/k)\, N_{e1} = 0 \end{cases}$$

It has non-trivial solution when

$$\left[\varepsilon + e^2/(c - \rho_m\omega^2/k^2)\right] \left[\mu\mathscr{E}_o + ikD_e + \omega/k\right] = N_{eo}e\mu/ik \tag{4}$$

Let $k = \omega/v_s + \delta$ where $v_s = \sqrt{c/\rho_m}$ and assume that $\delta \ll \omega/v_s$. Eqn (4) reduces to

$$\frac{e^2\omega}{2cv_s\delta}\left(\mu\mathscr{E}_o + iD_e\omega/v_s + v_s\right) = N_{eo}\mu ev_s/i\omega$$

or $\delta = e\omega^2 \left[- D_e\omega/v_s + i(\mu\mathscr{E}_o + v_s)\right] / 2N_{eo}\mu v_s^2 c$

There is gain when $\mathrm{Im}\ \delta < 0$

i.e. when $\mu\mathscr{E}_o + v_s < 0$

or $v_s < -\mu\mathscr{E}_o = v_o$ (Q.E.D.)

10.10 The differential equation describing the motion of electron is

$$m\frac{d^2x}{dt^2} + \gamma\frac{dx}{dt} + kx = e\mathscr{E}' \qquad \text{where } \mathscr{E}' \text{ is the local electric field.}$$

Take the Fourier transform to get the frequency response,

$$(-m\omega^2 - i\omega\gamma + k)\,X = e\mathscr{E}'$$

$$\therefore \text{ dipole moment } \mu = eX = \frac{\mathscr{E}'e^2/m}{\omega_0^2 - \omega^2 - \frac{i\omega\gamma}{m}} \qquad \text{where } \omega_0 = \sqrt{k/m}$$

But from [eqns 10.8 and 10.33], $\mu = \dfrac{\varepsilon_r - 1}{\varepsilon_r + 2}\dfrac{3\varepsilon_0}{N}\mathscr{E}'$

$$\therefore \frac{\varepsilon_r - 1}{\varepsilon_r + 2} = \frac{Ne^2/3m\varepsilon_0}{\omega_0^2 - \omega^2 - \frac{i\omega\gamma}{m}}$$

$$\text{giving} \quad \varepsilon_r = 1 + \frac{Ne^2/m\varepsilon_0}{\omega_0^2 - \omega^2 - \frac{i\omega\gamma}{m} - Ne^2/3m\varepsilon_0} = \varepsilon_r' + i\varepsilon_r''$$

where

$$\begin{cases} \varepsilon_r' = 1 + \dfrac{Ne^2}{m\varepsilon_0}\left[\dfrac{\omega_1^2 - \omega^2}{(\omega_1^2 - \omega^2)^2 + (\omega\gamma/m)^2}\right] \\[2ex] \varepsilon_r'' = \dfrac{Ne^2}{m\varepsilon_0}\left[\dfrac{\omega\gamma/m}{(\omega_1^2 - \omega^2)^2 + (\omega\gamma/m)^2}\right] \end{cases}$$

and $\omega_1^2 = \omega_0^2 - Ne^2/3m\varepsilon_0$

10.11 The relative permittivity will relax to a low value at its natural frequency. From [eqn 5.6], the equation of motion of the atoms can be written as

$$M\frac{d^2(\Delta r)}{dt^2} = -T\,r_0^2 = -3cr_0\,\Delta r$$

where M is the mass of the atom (take the average mass of K and Cl atoms)

Thus the natural frequency is $\sqrt{3cr_o/M}\,/2\pi$. Now $c = 1.88\text{x}10^{10}$ Nm^{-2}, r_o = $3.12\text{x}10^{-10}$ m (from Ex 5.4), $M = (39.1 + 35.5)/(2\text{x}6.02\text{x}10^{26})$ kg,

∴ the natural frequency = <u>$2.68\text{x}10^{12}$ Hz</u>

11. Magnetic materials

11.1 [eqn 11.11] : $\chi_m = -\dfrac{N_a Ze^2r^2\mu_o}{4m}$

Take mass, length, time, current as the basic unit (denoted By M, L, T, I respectively).

Now H (Henry) = WbA^{-1}, Wb = Vs, V = $Js^{-1}A^{-1}$, J = kgm^2s^{-2}

$\therefore$ unit of μ_o = Hm^{-1} = $kgm^2s^{-2}\cdot s^{-1}A^{-1}\cdot s\cdot A^{-1}\cdot m^{-1}$ = $kgms^{-2}A^{-2}$

Hence the dimension of $N_a Ze^2r^2\mu_o/4m$ is (note that Z is dimensionless)

$$\frac{(L^{-3})\cdot(IT)^2\cdot(L)^2\cdot(MLT^{-2}I^{-2})}{M} = 1$$

Indeed χ_m is dimensionless. Thus [eqn 11.11] is dimensionally correct.

Take $N_a = 10^{28}\ m^{-3}$, z = 10, r = 10^{-10} m, we get $\chi_m = 8.8\text{x}10^{-6}$.

11.2 The induced magnetic moment is given by [eqn 11.9] :

$$(\mu_m)_{ind} = Br^2e^2/4m$$

With B = 1 Wbm^{-2}, r = 0.0528 nm (from [eqn 4.24]), e = $1.6\text{x}10^{-19}$ C, and m = $9.11\text{x}10^{-31}$ kg,

$$(\mu_m)_{ind} = \underline{1.96\text{x}10^{-29}\ Am^2}$$

1 Bohr magneton = $9.27\text{x}10^{-24}\ Am^2$ which is much larger than $(\mu_m)_{ind}$.

11.3 (i) The force on a wire of length ℓ carrying a current I under a magnetic field B is ℓ**IxB**. So for the rectangular current loop (dimension

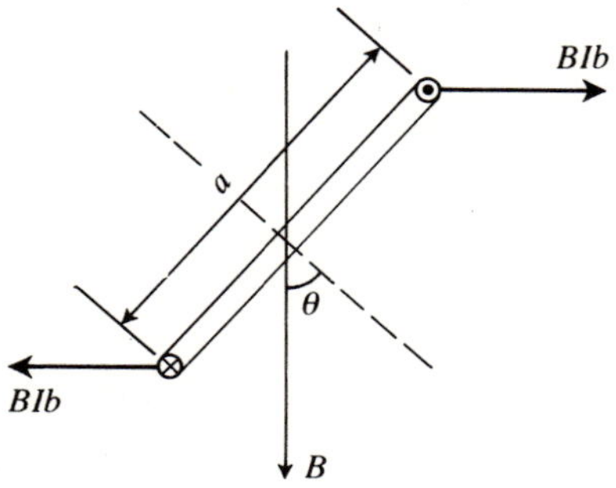

Fig 11.1 The forces on a rectangular current loop due to a magnetic field.

of axb), the net force is zero, but there is a net torque T = BIab sin θ (see Fig 11.1).

∴ the energy of the loop at an angle θ is

$$E = -\int BIab \sin\theta \, d\theta$$

$$= BIab \cos\theta \quad \text{(taking the energy to be zero at } \theta = \pi/2)$$

(ii) Now $E = \boldsymbol{\mu}_m \cdot \mathbf{B}$. Compare with the expression in part (i), the magnetic moment vector of the loop is

$\boldsymbol{\mu}_m$ = area of the loop x I $\hat{\mathbf{n}}$ where $\hat{\mathbf{n}}$ is a unit vector normal to the loop.

(iii) At stable equilibrium, $\frac{dE}{d\theta} = 0$ and $\frac{d^2E}{d\theta^2} < 0$,

i.e. when θ = 0

In this case the magnetic field of the loop is in the direction of the applied field.

11.5 From [eqn 11.26], $\lambda = \theta/C$ and $\mu_m = (3kC/N_m\mu_o)^{1/2}$

From experiments, $\theta = 1043$ K and $C \simeq 1$ for iron.

Take $N_m = 8x10^{28}\ m^{-3}$,

$\therefore \lambda = 1043/1 \simeq \underline{1000}$

and $\mu_m = \left[\ 3x1.38x10^{-23}x1/(4\pi x10^{-7}x8x10^{28})\ \right]^{1/2} \simeq \underline{2x10^{-23}\ Am^2}$

11.6 Curie law [eqn 11.25] is $\chi_m = C/(T - \theta)$

Rearrange to $T = C\chi_m^{-1} + \theta$

Plot T Vs χ_m^{-1} (see Fig 11.2)

Slope of curve = C (Curie constant) = $\underline{4.98x10^{-2}\ K}$

By extrapolation, at $\chi_m^{-1} = 0$, $T = \theta$ (Curie temperature) = $\underline{633\ K}$

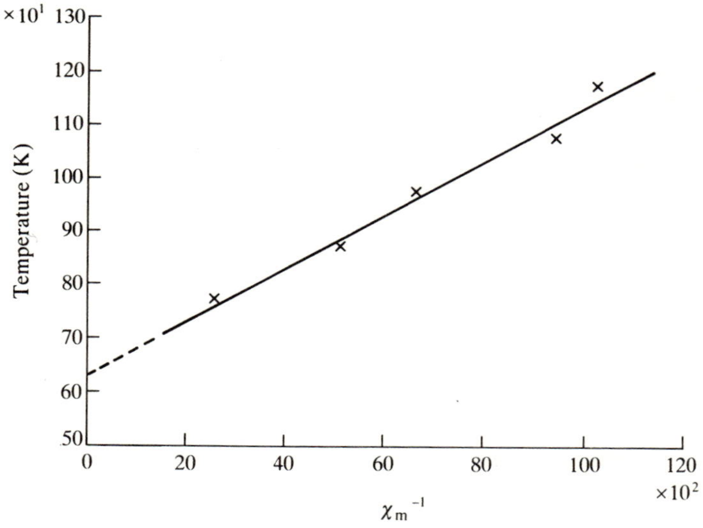

Fig 11.2 Plot of temperature T as a function of χ_m^{-1}.

From [eqn 11.24], $\mu_m = (3kC/\mu_o N_m)^{1/2}$

$$= \left[3x1.38x10^{-23}x4.98x10^{-2}/(4\pi x10^{-7}x6.02x10^{26}x8850/58.7) \right]^{1/2}$$

$$= 4.25x10^{-24} \text{ Am}^2$$

∴ the effective number of Bohr magnetons per atom is

$4.25x10^{-24}/9.27x10^{-24} = \underline{0.46}$

11.7 From [eqns 11.15 and 11.16], $\chi_m = M_s^2\mu_o/3kTN_m$

Now N_m = density of the precipitates (magnetic dipole)

$$= 0.02 / \left[\frac{4}{3} \pi (5x10^{-9})^3 \right] = 3.82x10^{22} \text{ m}^{-3}$$

and M_s = the saturation magnetisation of the alloy

$$= 0.02x1.4x10^6 = 2.8x10^4 \text{ Am}^{-1}$$

∴ $\chi_m = (2.8x10^4)^2x4\pi x10^{-7} / (3x1.38x10^{-23}x300x3.82x10^{22}) = \underline{2.08}$

11.8 From the Boltzmann statistics, $N_i = N_o \exp(-E_i/kT)$ where i = 1, 2 corresponding to the parallel and antiparallel spins respectively.

Now the energy difference between the two states = $2\mu_{mB}B$

∴ $N_1/N_2 = \exp 2\mu_{mB}B/kT$

With $N_1/N_2 = 2$, $\mu_{mB} = 9.27x10^{-24}$ Am^2 and B = 2 Wbm^{-2}, we get T = $\underline{3.88 \text{ K}}$

11.9 There is magnetic resonance when the frequency f is given by

$f = \frac{1}{2\pi}\frac{ge}{2m}B$ (from [eqn 11.48]) where the constant g = 2, 5.58 for electron spin resonance and proton spin resonance respectively.

Hence (i) $f_e = \frac{1}{2\pi} \times \frac{2 \times 1.6 \times 10^{-19}}{2 \times 9.11 \times 10^{-31}} \times 0.1 = \underline{2.8 \times 10^9 \text{ Hz}}$

(ii) $f_p = \frac{1}{2\pi} \times \frac{5.58 \times 1.6 \times 10^{-19}}{2 \times 1.67 \times 10^{-27}} \times 0.1 = \underline{4.3 \times 10^6 \text{ Hz}}$

11.10 The magnetic moment due to an 'up'-spin electron will cancel with a 'down'-spin electron. Therefore the net magnetic moment is equal to the difference between the number of 'up'-spin and 'down'-spin electrons (denoted by ΔN) multiplied by the magnetic moment due to each electron.

Now $\Delta N = \frac{1}{2}\int_0^{E_F} Z(E + \mu_m\mu_o H)\, dE - \frac{1}{2}\int_0^{E_F} Z(E - \mu_m\mu_o H)\, dE$

$$= \frac{1}{2}\int_0^{E_F}\left[Z(E) + \mu_m\mu_o H \frac{dZ}{dE}\right] dE - \frac{1}{2}\int_0^{E_F}\left[Z(E) - \mu_m\mu_o H \frac{dZ}{dE}\right] dE$$

$$= \mu_m\mu_o H \int_0^{E_F} \frac{dZ}{dE}\, dE = \mu_m\mu_o H\, Z(E_F)$$

Hence the total magnetic moment $M = \mu_m^2\mu_o H\, Z(E_F)$

$\therefore \chi_m = M/H = \mu_m^2\mu_o Z(E_F)$ (Q.E.D.)

12. Lasers

12.1 (a) From [eqn 12.11], the ratio of the Einstein coefficients is

$$A/B = 8\pi n^3 h\nu^3/c^3 = 8\pi n^3 h/\lambda^3$$

$= \underline{5.0\times10^{-14}}$ for $\lambda = 693$ nm; $\underline{4.9\times10^{-27}}$ for $\lambda = 1.5$ cm.

(b) The ratio of spontaneous transitions to stimulated transitions

$= A/B\rho(\nu)$ where $\rho(\nu)$ is the radiation density at frequency ν

$= \exp h\nu/kT - 1$ (from [eqn 12.9] where $B_{13} = B_{31} = B$)

$= \underline{1.16\times10^{30}}$ for $\lambda = 693$ nm, $T = 300$ K; $\underline{0.27}$ for $\lambda = 1.5$ cm, $T = 4$ K.

This ratio is equal to unity when $\exp h\nu/kT - 1 = 1$

At room temperature (293 K), this gives $\nu = \underline{4.23\times10^{12}\ \text{Hz}}$.

12.3 From the Boltzmann's statistics, $N = N_o \exp(-E/kT)$ [eqn 12.1].

For hydrogen, the energy difference between the ground state and the excited state (n= 2) is 13.6x(1 - 1/4) = 10.2 eV. Therefore the fraction of atoms in the excited state at 3500 K is

$$N/N_o = \exp(-10.2/8.62\times10^{-5}\times3500) = \underline{2.08\times10^{-15}}$$

The number of atoms returning to the ground state in 1 second is

$$10^{21}\times2.08\times10^{-15}/10^{-8}$$

Each transition will liberate 10.2 eV.

$\therefore$ the total radiated power $= 10.2\times10^{21}\times2.08\times10^{-15}/10^{-8}$ eV $= \underline{339\ \mu W}$

The frequency of radiation $= 10.2\text{x}1.6\text{x}10^{-19}/6.62\text{x}10^{-34} = 2.47\text{x}10^{15}$ Hz (wavelength = 122 nm) which is not in the visible range.

12.4 From [eqn 12.4], the excess of upper level population density over lower population density is

$$N_3 - N_2 = 8\pi n^2\nu^2 t_{spont}\gamma(\nu)/c^2 g(\nu) = 8\pi n^2 t_{spont}\gamma(\nu)/\lambda^2 g(\nu)$$

$$= \frac{8\pi\text{x}1.77^2\text{x}3\text{x}10^{-3}\text{x}0.04}{(693\text{x}10^{-9})^2/2\text{x}10^{11}} \quad \text{(note that linewidth } \Delta\nu = 1/g(\nu_o))$$

$$= \underline{3.93\text{x}10^{21}\ \text{m}^{-3}}$$

12.5 In order that laser action is possible, the loop gain must be larger than unity, i.e. $R_1R_2 \exp 2\ell(\gamma - \alpha) \geq 1$ ([eqn 12.33]) where R_1, R_2 are the reflectivities of the mirrors, γ and α are the gain and attenuation coefficients respectively.

∴ the minimum reflectivity of the second mirror is

$$\left[\exp 2\text{x}3\text{x}(0.148 - 0.14)\right]^{-1} = \underline{0.953}$$

12.6 Input power = VI = 500x50 W = 25 kW

∴ the efficiency of the laser $= 5/25\text{x}10^3 = \underline{2\text{x}10^{-4}}$

12.7 The lasing wavelength must satisfied

$2\ell = n\lambda$ where n is an integer, ℓ is the length of the resonator and λ is the wavelength in the medium.

$\therefore\ \nu = nv/2\ell$ where v is the speed of the wave in the resonator.

Thus the frequency difference between the nearest modes is

$nv/2\ell - (n-1)v/2\ell = v/2\ell$

12.8 (i) The power is half the maximum when (refer to [eqn 12.26])

$$\exp\left[-\,Mc^2(\nu - \nu_o)^2/2kT\nu_o^2\right] = \frac{1}{2}$$

or $$\nu - \nu_o = \pm\,\nu_o\left[\,2kT\ \ln 2\ /\ Mc^2\,\right]^{1/2}$$

Thus, the half-power bandwidth is

$$\Delta\nu = 2\nu_o\left[\,2kT\ \ln 2\ /\ Mc^2\,\right]^{1/2} \qquad \text{(Q.E.D.)}$$

(ii) For 514.5 nm, 5000 K,

$$\Delta\nu = \frac{2}{514.5\text{x}10^{-9}}\left(\frac{2\text{x}1.38\text{x}10^{-23}\text{x}5000\text{x}\ln 2}{40/6.02\text{x}10^{26}}\right)^{1/2} = \underline{4.66\text{x}10^{9}\ \text{Hz}}$$

(iii) From the result of Ex 12.7, the frequency difference between the nearest mode is $v/2\ell$ where v is the speed of the wave in the resonator medium and ℓ is the length of the resonator. So for $v = 3\text{x}10^8\ \text{ms}^{-1}$ and $\ell = 1.5$ m, $v/2\ell = 10^8$ Hz. Now $\Delta\nu = 4.66\text{x}10^9$ Hz, therefore 46 different modes are possible.

12.10 The threshold current density J is given by [eqn 12.38]

$$J = \frac{8\pi n^2\nu^2 ed}{c^2\eta g(\nu)}\left(\alpha - \frac{1}{\ell}\ln R\right)$$

Now the reflectivity is due to a cleaved edge (a change in refractive index from 3.35 to 1).

$$\therefore R = \left(\frac{3.35 - 1}{3.35 + 1}\right)^2 = 0.292$$

$$\therefore J = \frac{8\pi \times 3.35^2 \times 1.6\times10^{-19} \times 2\times10^{-6}}{(887\times10^{-9})^2 \times 1/10^{13}} \left(10^3 - \frac{\ln 0.292}{0.2\times10^{-3}}\right) = \underline{8.2\times10^6\ \mathrm{Am}^{-2}}$$

12.11 From the definition of Q,

$$\text{rate of energy lost} = \frac{\omega \times \text{energy stored}}{Q} = \frac{\omega\mu_o H^2 V}{Q} \qquad \text{(Q.E.D.)}$$

The power of stimulated emission is

$$V\hbar\omega\ \Delta N\ \alpha H^2$$

The condition for maser oscillation is that the power gain is larger than the power lost,

i.e. $V\hbar\omega\ \Delta N\ \alpha H^2 > \omega\mu_o H^2 V/Q$

giving $\Delta N > \mu_o/\alpha\hbar Q$ (Q.E.D.)

13. Optoelectronics

13.1 At a depth x, the intensity of light = $I_o \exp(-\alpha x)$.

Each photon has an energy of $h\nu$ and each photon will induce η electrons, each of which has a life time τ_e.

$\therefore$ the excess density of electrons at a depth x is $I_o \exp(-\alpha x)\ \eta\tau_e/h\nu$.

Now the mobility is μ_e and the electric field is V/c,

$$\therefore \text{ the excess current } \Delta I = \int_0^d - \frac{\eta\tau_e I_o \exp(-\alpha x)}{h\nu}\ e\mu_e \frac{V}{c}\ b\ dx$$

$$= e \frac{b}{c} \frac{\eta I_o}{h\nu} \tau_e\mu_e V \frac{1 - e^{-\alpha d}}{\alpha} \qquad \text{(Q.E.D.)}$$

13.2 No. of photocarriers crossing the electrodes per unit time is

$$\int_0^d \frac{\eta I_o \exp(-\alpha x)}{h\nu} \tau_e\mu_e \frac{V}{c}\ b\ dx$$

No. of photocarriers generated per unit time is $\displaystyle\int_0^d \frac{\eta I_o \exp(-\alpha x)}{h\nu}\ bc\ dx$

$\therefore$ the photoconductive gain $G = \underline{\tau_e\mu_e V/c^2}$

13.3 Assume that all the donors of the lightly doped n-type region are ionized. From Maxwell's equation,

$$\varepsilon \frac{d\mathcal{E}}{dx} = eN_{D1}$$

$$\therefore \mathcal{E}(x) = eN_{D1}x/\varepsilon + \mathcal{E}(0)$$

Further assume that the depletion region is all in this region. Then the voltage across the diode is

$$V = -\int_0^{d_2} \mathcal{E}(x)\, dx \quad \text{(taking the } p^+\text{-n boundary as } x = 0)$$

$$= -\left[eN_{D1}d_2^2/2\varepsilon + \mathcal{E}(0)d_2\right]$$

$$\therefore \mathcal{E}(0) = -\left[eN_{D1}d_2/2\varepsilon + V/d_2\right]$$

Hence the electric field across the junction looks like Fig 13.1.

The p^+ region must be extremely thin so that the incident light can penetrate to the junction to create the carriers.

13.4 (i) From [eqn 13.8], the grating spacing is given by

$$\Lambda = \frac{\lambda}{2n \sin\theta} = \frac{514.5}{2 \times 1.52 \times \sin 5}\ \text{nm} = \underline{1.94\ \mu\text{m}}$$

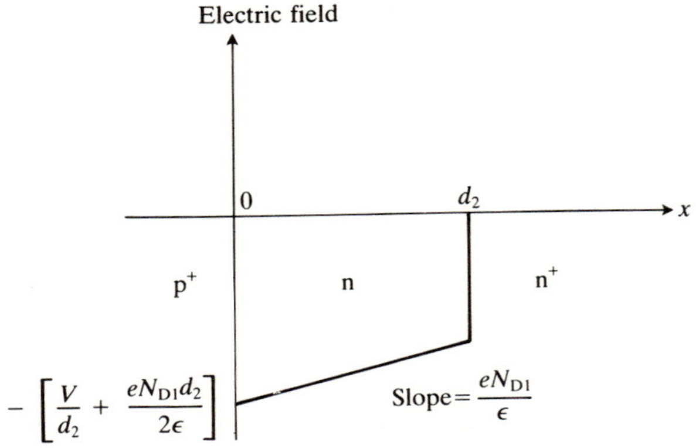

Fig 13.1 The electric field variation across the diode.

(ii) The Bragg angle may be calculated from the same formula but now λ = 633 nm. Therefore $\frac{633\text{x}10^{-9}}{2\text{x}1.52\text{x}\sin\theta} = 1.94\text{x}10^{-6}$, giving $\theta = \underline{6.16^{\circ}}$

(iii) When replaying at a wavelength λ, the second Bragg angle θ is given by $\frac{\lambda}{2n\sin\theta} = \frac{\Lambda}{2}$

So for λ = 514.5 nm, $\theta = \underline{10.03^{\circ}}$ and for λ = 633 nm, $\theta = \underline{12.40^{\circ}}$

13.5 (i) The two incident waves are A_{10} exp ikx and A_{20} exp(−ikx).

$\therefore$ the intensity $\propto |A_{10} \exp ikx + A_{20} \exp(-ikx)|^2$

$$= A_{10}^2 + A_{20}^2 + 2A_{10}A_{20}\cos 2kx$$

$\therefore$ the grating spacing $\Lambda = \frac{2\pi}{2k} = \frac{\lambda}{2n}$

Hence at 514.5 nm, Λ = <u>169.2 nm</u> and at 633 nm, Λ = <u>208.2 nm</u>.

(ii) For the same incident angle, the grating spacing is much smaller in a reflection type than in a transmission type. Thus the former is more sensitive to the replay wavelength.

13.6 (i) When an electric field $\mathscr{E}_Z$ is applied in the z-direction, the electric tensor becomes

$$\varepsilon = \varepsilon_0 \begin{bmatrix} \varepsilon_r & -\varepsilon_r^2 r_{XYZ}\mathscr{E}_Z & 0 \\ -\varepsilon_r^2 r_{XYZ}\mathscr{E}_Z & \varepsilon_r & 0 \\ 0 & 0 & \varepsilon_r \end{bmatrix}$$

(ii) Consider $\mathbf{D}_{XYZ} = \varepsilon \mathscr{E}_{XYZ}$

The vectors $\mathbf{D}_{XYZ}$, $\mathscr{E}_{XYZ}$ can be transformed to xyz coordinate system by

$$\mathbf{D}_{xyz} = C\mathbf{D}_{XYZ} \text{ and } \mathscr{E}_{xyz} = C\mathscr{E}_{XYZ}$$

$$\text{where } C = \begin{bmatrix} \cos\theta & -\sin\theta & 0 \\ \sin\theta & \cos\theta & 0 \\ 0 & 0 & 1 \end{bmatrix}$$

$$\text{Thus } C^{-1}\mathbf{D}_{xyz} = \varepsilon C^{-1}\mathscr{E}_{xyz}$$

$$\text{or } \mathbf{D}_{xyz} = C\varepsilon C^{-1}\mathscr{E}_{xyz}$$

∴ in xyz coordinate, the dielectric tensor becomes $C\varepsilon C^{-1}$,

$$\text{i.e. } \varepsilon_o \begin{bmatrix} \varepsilon_r + \varepsilon_r^2 r_{XYZ}\mathscr{E}_Z \sin 2\theta & -\varepsilon_r^2 r_{XYZ}\mathscr{E}_Z \cos 2\theta & 0 \\ -\varepsilon_r^2 r_{XYZ}\mathscr{E}_Z \cos 2\theta & \varepsilon_r - \varepsilon_r^2 r_{XYZ}\mathscr{E}_Z \sin 2\theta & 0 \\ 0 & 0 & \varepsilon_r \end{bmatrix}$$

(iii) When $\theta = 45^\circ$, it becomes

$$\varepsilon_o \begin{bmatrix} \varepsilon_r + \varepsilon_r^2 r_{XYZ}\mathscr{E}_Z & 0 & 0 \\ 0 & \varepsilon_r - \varepsilon_r^2 r_{XYZ}\mathscr{E}_Z & 0 \\ 0 & 0 & \varepsilon_r \end{bmatrix}$$

If the input wave is polarized in the y-direction then the relevant element of the dielectric tensor ε_{yy} may be seen to depend on the applied electric field.

13.7 (i) Eliminate A_1 from [eqns 13.16 and 13.17],

$$\frac{d^2A_2}{dz^2} - i\left(k_1 + k_2\right)\frac{dA_2}{dz} + \left(\kappa^2 - k_1k_2\right)A_2 = 0$$

$\therefore\ A_2 = \alpha \exp \gamma_1 z + \beta \exp \gamma_2 z$ where α, β are constants, (1)

and $\gamma_{1,2} = i(k_1 + k_2)/2 \pm i\phi$, $\phi = \left[\,(k_1 - k_2)^2/4 + \kappa^2\,\right]^{1/2}$

Substitute into [eqn 13.17] and rearrange to get

$$A_1 = \frac{\alpha(\gamma_1 - ik_2)}{i\kappa}\exp \gamma_1 z + \frac{\beta(\gamma_2 - ik_2)}{i\kappa}\exp \gamma_2 z \qquad (2)$$

At $z = 0$, $A_1 = A_{10}$ and $A_2 = 0$.

$\therefore\ \alpha + \beta = 0$ and $\alpha(\gamma_1 - ik_2) + \beta(\gamma_2 - ik_2) = i\kappa A_{10}$

yielding $\alpha = -\beta = i\kappa A_{10}/(\gamma_1 - \gamma_2) = 2\kappa A_{10}/\phi$

Substitute into eqns (1) and (2), and simplify to get,

$$\begin{cases} A_1 = A_{10}\exp\left[i(k_1 + k_2)z/2\right]\left[\cos \phi z + i\left((k_1 - k_2)/2\phi\right)\sin \phi z\right] & (3) \\ A_2 = \left(i\kappa A_{10}/\phi\right)\exp\left[i(k_1 + k_2)z/2\right]\sin \phi z & (4) \end{cases}$$

(ii) When $k_1 = k_2 = k$, eqns (3) and (4) reduce to

$A_1 = A_{10}\exp(ikz)\cos \kappa z$ and $A_2 = iA_{10}\exp(ikz)\sin \kappa z$ (Q.E.D.)

(iii) There will be complete power transfer when $\kappa L = \pi/2$.

So for L = 1 cm, $\kappa = \underline{\pi/2\ \text{cm}^{-1}}$

(iv) The maximum power coupled to waveguide 2 is $|A_{10}|^2$ when $\kappa = \phi = \pi/2\ \text{cm}^{-1}$. When $v_1 \neq v_2$, then $\phi \neq \kappa$ and the equation to satisfy is

$$|A_2(L)|^2 = |A_{10}|^2 \frac{\pi^2}{4} \frac{\sin^2\phi L}{\phi^2} = \frac{1}{2} |A_{10}|^2$$

whence $\frac{\sin \phi L}{\phi} = \sqrt{2}/\pi$

Solving numerically to get $\phi = 2.01\ cm^{-1}$

$$\therefore\ \Delta k = k_1 - k_2 = 2\left(\phi^2 - \kappa^2\right)^{1/2} = \pm\ 2.508\ cm^{-1}$$

But $k = \omega/v$, so $\Delta k = -\omega\Delta v/v^2$.

$$\therefore\ \Delta v = -v^2\Delta k/\omega = \pm\ 10^{16}x2.508x10^2/(2\pi x3x10^8/633x10^{-9}) = \underline{\pm\ 842.2\ ms^{-1}}$$

13.8 The transmitted electric field is given by

$$\mathscr{E}_t = \mathscr{E}_i t_1 t_2 e^{(\gamma/2\ +\ ik)L} + \mathscr{E}_i t_1 r_2 r_1 t_2 e^{2(\gamma/2\ +\ ik)L} + \mathscr{E}_i t_1 r_2 r_1 r_2 r_1 t_2 e^{3(\gamma/2\ +\ ik)L} + \cdots$$

$$= \mathscr{E}_i t_1 t_2 e^{(\gamma/2\ +\ ik)L} \left\{ 1 + r_1 r_2 e^{(\gamma/2\ +\ ik)L} + \left[r_1 r_2 e^{(\gamma/2\ +\ ik)L} \right]^2 + \cdots \right\}$$

$$= \frac{t_1 t_2\ \exp(\gamma/2\ +\ ik)L}{1 - r_1 r_2\ \exp(\gamma/2\ +\ ik)L} \mathscr{E}_i \qquad \text{assuming that } \left| r_1 r_2 e^{\gamma L/2} \right| < 1$$

14. Superconductivity

14.1 Due to the accuracy of the measurement, the current may have declined to $J_o(1 - 10^{-4})$. According to the equation $J = J_o\exp(-t/\tau)$, this would imply the decay constant τ to be $-t/\log(1 - 10^{-4})$ where t is one year, giving $\tau \simeq \underline{3.15\text{x}10^{11}\ \text{s}}$

From [eqn 1.10], $\sigma = Ne^2\tau/m = 10^{28}\text{x}(1.6\text{x}10^{-19})^2\text{x}3.15\ 10^{11}/9.11\text{x}10^{-31}$

$= \underline{8.85\text{x}10^{31}\ \Omega^{-1}\text{m}^{-1}}$

The conductivity of copper is about $5.8\text{x}10^{7}\ \Omega^{-1}\text{m}^{-1}$.

Hence $\sigma/\sigma_{Cu} \simeq \underline{1.5\text{x}10^{24}}$

14.2 The critical magnetic field for the superconducting state is given by [eqn 14.1] : $H_c = H_o\left[1 - (T/T_c)^2\right]$

From [Table 14.1], $T_c = 7.18$ K, $H_o = 6.5\text{x}10^{-4}\ \text{Am}^{-1}$ for lead.

$\therefore\ H_c = 6.5\text{x}10^{-4}\left[1 - (5/7.18)^2\right] = 3.35\text{x}10^{-4}\ \text{Am}^{-1}$ at 5 K

From Ampere's law the relationship between the current and the magnetic field just outside the wire is $I = H/2\pi r = 3.35\text{x}10^{-4}/(2\pi\text{x}10^{-3}) = \underline{0.053\ \text{A}}$

Note that this calculation is not valid for type II superconductors when the magnetic field penetrates the superconductor.

14.3 From Maxwell's equations, $\nabla\cdot\mathbf{H} = \mathbf{J}$

But from London's assumption $\mathbf{J} = -\alpha\mathbf{A}$ where α is a proportionality constant.

$$\therefore\ -\alpha\mathbf{A} = \frac{1}{\mu_o}\nabla\times\mathbf{B} = \frac{1}{\mu_o}\nabla\times(\nabla\times\mathbf{A}) \quad \text{(by the def. of vector potential A)}$$

$$= \frac{1}{\mu_o}\left[\nabla(\nabla\cdot\mathbf{A} - \nabla^2\mathbf{A})\right] = -\frac{1}{\mu_o}\nabla^2\mathbf{A} \quad (\because \nabla\cdot\mathbf{A} = 0)$$

or $\frac{d^2A}{dx^2} - \alpha\mu_o A = 0$ in one dimension (Q.E.D.)

14.4 From [eqn 14.60],

$$\psi_o^2 = \frac{m}{4e^2\mu_o\lambda^2} = \frac{9.11\times10^{-31}}{4\times(1.6\times10^{-19})^2\times4\pi\times10^{-7}\times(60\times10^{-9})^2} = \underline{1.97\times10^{27}\ \mathrm{m}^{-3}}$$

14.5 At absolute zero, when a voltage U is applied, the energy states are filled up to $E_F + eU$ in the 'left' superconductor. In the 'right' superconductor, the energy states are empty above E_F. The tunnelling current in each energy range dE is proportional to the number of filled states in the 'left' superconductor and to the number of empty states in the 'right' superconductor. However, the densities of states are zero between the band gaps. Therefore the density of filled states in the 'left' superconductor is (measure the energy from the Fermi level of the right superconductor)

$$\frac{c(eU - E)}{[(eU - E)^2 - \Delta^2]^{1/2}} \quad \text{for } E < eU - \Delta \text{ and zero otherwise}$$

and in the 'right superconductor, density of empty states is

$$\frac{cE}{(E^2 - \Delta^2)^{1/2}} \quad \text{for } E > \Delta \text{ and zero otherwise.}$$

Thus the tunnelling current is zero if $eU < 2\Delta$ and when $eU > 2\Delta$, the tunnelling current is proportional to

$$\int_{\Delta}^{eU-\Delta} \frac{eU - E}{[(eU - E)^2 - \Delta^2]^{1/2}} \frac{E}{(E^2 - \Delta^2)^{1/2}} dE \qquad \text{(Q.E.D.)}$$

14.6 (i) From [Fig 14.17], the tunnelling current is maximum when $U = (\Delta_2 - \Delta_1)/e$ and is at the point of upsurge when $U = (\Delta_1 + \Delta_2)/e$.

Hence $\Delta_2 - \Delta_1 = 0.52$ meV and $\Delta_1 + \Delta_2 = 1.65$ meV.

Solving to get the energy gaps ($2\Delta_1$ and $2\Delta_2$) to be 1.13 and 2.17 meV.

(ii) The current maximum will disappear when the temperature is raised above the critical temperature of the superconductor with the lower critical temperature, i.e. 3.72 K for tin (from [Table 14.1]).

14.7 The critical temperature of tin is 3.72 K (from [Table 14.1]). Thus the losses will be substantially less if the cavity is cooled from 4 K (normal phase) to 1 K at which the conductivity drops significantly (superconducting phase).

At 1 K, the energy gap of tin is about 2.17meV (from Ex 14.6). The superconductive effects will completely disappear when hf = 2.17 meV, i.e. when the frequency $f = 5.24 \times 10^{11}$ Hz.

14.8 From [eqn 14.73], $2eU_{AB} = h\omega = hf$

∴ freq. of radiation, $f = 2eU_{AB}/h = \dfrac{2 \times 1.6 \times 10^{-19} \times 650 \times 10^{-6}}{6.62 \times 10^{-34}} = 3.14 \times 10^{11}$ Hz